Using **R** for Data Management, Statistical Analysis, and Graphics

Using R for Data Management, Statistical Analysis, and Graphics

Nicholas J. Horton

Department of Mathematics and Statistics

Smith College

Northampton, Massachusetts

Ken Kleinman

Department of Population Medicine

Harvard Medical School

Boston, Massachusetts

CRC Press
Taylor & Francis Group
Boca Raton   London   New York

CRC Press is an imprint of the
Taylor & Francis Group an **informa** business

A CHAPMAN & HALL BOOK

CRC Press
Taylor & Francis Group
6000 Broken Sound Parkway NW, Suite 300
Boca Raton, FL 33487-2742

© 2011 by Taylor and Francis Group, LLC
CRC Press is an imprint of Taylor & Francis Group, an Informa business

No claim to original U.S. Government works

Printed in the United States of America on acid-free paper
10 9 8 7 6 5 4 3 2 1

International Standard Book Number: 978-1-4398-2755-0 (Paperback)

Library of Congress Cataloging-in-Publication Data

Horton, Nicholas J.
    Using R for data management, statistical analysis, and graphics / Nick Horton, Ken Kleinman.
        p. cm.
    Includes bibliographical references and index.
    ISBN 978 1 4398 2755 0 (pbk. : alk. paper)
    1. R (Computer program language) 2. Open-source software. 3. Database management. 4. Mathematical statistics--Data processing. I. Kleinman, Ken. II. Title.

QA76.73.R3H67 2011
519.50285'5133--dc22   2010021409

Visit the Taylor & Francis Web site at
http://www.taylorandfrancis.com

and the CRC Press Web site at
http://www.crcpress.com

Contents

List of Tables

List of Figures

Preface

R is a general purpose statistical software package used in many fields of research. It is licensed for free, as open-source software. The system has been developed by a large group of people, almost all volunteers. It has a large and growing user and developer base. Methodologists often release applications for general use in R shortly after they have been introduced into the literature. While professional customer support is not provided, there are many resources to help support users.

We have written this book as a reference text for users of R. Our primary goal is to provide users with an easy way to learn how to perform an analytic task in this system, without having to navigate through the extensive, idiosyncratic, and sometimes unwieldy documentation or to sort through the huge number of add-on packages. We include many common tasks, including data management, descriptive summaries, inferential procedures, regression analysis, multivariate methods, and the creation of graphics. We also show some more complex applications. In toto, we hope that the text will facilitate more efficient use of this powerful system.

We do not attempt to exhaustively detail all possible ways available to accomplish a given task in each system. Neither do we claim to provide the most elegant solution. We have tried to provide a simple approach that is easy to understand for a new user, and have supplied several solutions when it seems likely to be helpful.

Who should use this book

Those with an understanding of statistics at the level of multiple-regression analysis should find this book helpful. This group includes professional analysts who use statistical packages almost every day as well as statisticians, epidemiologists, economists, engineers, physicians, sociologists, and others engaged in research or data analysis. We anticipate that this tool will be particularly useful for sophisticated users, those with years of experience in only one system, who need or want to use the other system. However, intermediate-level analysts should reap the same benefits. In addition, the book will bolster the analytic abilities of a relatively new user, by providing a concise reference manual and annotated examples.

Using the book

The book has two indices, in addition to the comprehensive Table of Contents. These include: (1) a detailed topic (subject) index in English and (2) an R command index, describing R syntax.

Extensive example analyses of data from a clinical trial are presented; see Table A.1 (in the Appendix) for a comprehensive list. These employ a single dataset (from the HELP study), described in the Appendix. Readers are encouraged to download the dataset and code from the book's Web site. The examples demonstrate the code in action and facilitate exploration by the reader.

In addition to the HELP examples, Chapter 7 features a varied set of case studies and extended examples that utilize many of the functions, idioms, and code samples from the earlier chapters. These include explications of analytic and empirical power calculations, missing data methods, propensity score analysis, sophisticated data manipulation, data gleaning from Web sites, map making, simulation studies, and optimization. Entries from earlier chapters are cross-referenced to help guide the reader.

Where to begin

We do not anticipate that the book will be read cover to cover. Instead, we hope that the extensive indexing, cross-referencing, and worked examples will make it possible for readers to directly find and then implement what they need. A new user should begin by reading the first chapter, which includes a sample session and overview of the system. Experienced users may find the case studies in Chapter 7 to be valuable as a source of ideas on problem solving in R.

On the Web

The book's Web site at `http://www.math.smith.edu/r` includes the Table of Contents, the Indices, the HELP dataset, example code, and a list of erratum.

Acknowledgments

We would like to thank Rob Calver, Kari Budyk, Linda Leggio, Shashi Kumar, and Sarah Morris for their support and guidance at Taylor & Francis CRC Press/Chapman & Hall. We also thank Ben Cowling, Stephanie Greenlaw, Tanya Hakim, Albyn Jones, Michael Lavine, Pamela Matheson, Elizabeth Stuart, Rebbecca Wilson, and Andrew Zieffler for comments, guidance, and helpful suggestions on drafts of the manuscript.

Above all we greatly appreciate Julia and Sara as well as Abby, Alana, Kinari, and Sam, for their patience and support.

Northampton and Amherst, Massachusetts

Chapter 1

Introduction to R

This chapter provides a (brief) introduction to R, a powerful and extensible free software environment for statistical computing and graphics [30, 49]. The chapter includes a short history, installation information, a sample session, background on fundamental structures and actions, information about help and documentation, and other important topics.

R is a general purpose package that includes support for a wide variety of modern statistical and graphical methods (many of which are included through user contributed packages). It is available for most UNIX platforms, Windows and Mac OS. R is part of the GNU project, and is distributed under a free software copyleft (`http://www.gnu.org/copyleft/gpl.html`). The R Foundation for Statistical Computing holds and administers the copyright of R software and documentation.

The first versions of R were written by Ross Ihaka and Robert Gentleman at the University of Auckland, New Zealand, while current development is coordinated by the R Development Core Team, a committed group of volunteers. As of February 2010 this consisted of Douglas Bates, John Chambers, Peter Dalgaard, Seth Falcon, Robert Gentleman, Kurt Hornik, Stefano Iacus, Ross Ihaka, Friedrich Leisch, Thomas Lumley, Martin Maechler, Duncan Murdoch, Paul Murrell, Martyn Plummer, Brian Ripley, Deepayan Sarkar, Duncan Temple Lang, Luke Tierney, and Simon Urbanek. Many hundreds of other people have contributed to the development or created add-on libraries and packages on a volunteer basis [20].

R is similar to the S language, a flexible and extensible statistical environment originally developed in the 1980s at AT&T Bell Labs (now Lucent Technologies). Insightful Corporation has continued the development of S in their commercial software package S-PLUS™.

New users are encouraged to download and install R from the Comprehensive R archive network (CRAN) (Section 1.1), then review this chapter. The sample session in the Appendix of the *Introduction to R* document, also available from CRAN (see Section 1.2), is also helpful to get an overview.

1

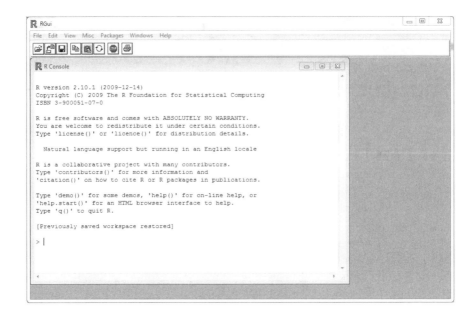

Figure 1.1: Windows graphical user interface.

1.1   Installation

The home page for the R project, located at http://r-project.org, is the best starting place for information about the software. It includes links to CRAN, which features precompiled binaries as well as source code for R, add-on packages, documentation (including manuals, frequently asked questions, and the R newsletter) as well as general background information. Mirrored CRAN sites with identical copies of these files exist all around the world. New versions are regularly posted on CRAN, which must be downloaded and installed.

1.1.1   Installation under Windows

Precompiled distributions of R for Windows are available at CRAN. Two versions of the executable are installed: Rgui.exe, which launches a self-contained windowing system that includes a command-line interface, and Rterm.exe which is suitable for batch or command-line use. A screenshot of the graphical user interface (GUI) can be found in Figure 1.1.

The GUI includes a mechanism to save and load the history of commands from within an interactive session (see also 2.7.4, history of commands).

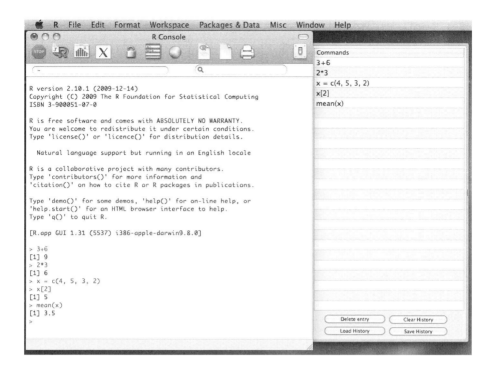

Figure 1.2: Mac OS X graphical user interface.

More information on Windows-specific issues can be found in the CRAN *R for Windows FAQ* (http://cran.r-project.org/bin/windows/base/rw-FAQ.html).

1.1.2 Installation under Mac OS X

A precompiled universal binary for Mac OS X 10.5 and higher is available at CRAN. This is distributed as a disk image containing the installer. In addition to the graphical interface version, a command line version (particularly useful for batch operations) can be run as the command R. A screenshot of the graphical interface can be found in Figure 1.2.

The GUI includes a mechanism to save and load the history of commands from within an interactive session (see also 2.7.4, history of commands). More information on Macintosh-specific issues can be found in the CRAN *R for Mac OS X FAQ* (http://cran.r-project.org/bin/macosx/RMacOSX-FAQ.html).

1.1.3   Installation under Linux

Precompiled distributions of R binaries are available for the Debian, Redhat (Fedora), Suse and Ubuntu Linux, and detailed information on installation can be found at CRAN. There is no built-in graphical user interface for Linux (but see 1.3 for the R Commander project [19]).

1.2   Running R and sample session

Once installation is complete, the recommended next step for a new user would be to start R and run a sample session. An example from the command line interface within Mac OS X is given in Figure 1.3.

The '>' character is the prompt, and commands are executed once the user presses the RETURN key. R can be used as a calculator (as seen from the first two commands on lines 1 through 4). New variables can be created (as on lines 5 and 8) using the = assignment operator. If a command generates output (as on lines 6 to 7 and 11 to 12), then it is printed on the screen, preceded by a number indicating place in the vector (this is particularly useful if output is longer than one line, e.g., lines 24 and 25). A dataframe, assigned the name ds, is read in on line 15, then summary statistics are calculated (lines 21 and 22) and individual observations are displayed (lines 23 and 25). The $ operator (line 16) allows direct access to objects within a dataframe. Alternatively the attach() command (line 20) can be used to make objects within a dataset available in the global workspace from that point forward.

It is important to note that R is case-sensitive, as demonstrated in the following example.

```
> x = 1:3
> X = seq(2, 4)
> x
[1] 1 2 3
> X
[1] 2 3 4
```

A useful sample session can be found in the Appendix A of *An Introduction to R* [76] (http://cran.r-project.org/doc/manuals/R-intro.pdf). New users to R will find it helpful to run the commands from that sample session.

1.2.1   Replicating examples from the book and sourcing commands

To help facilitate reproducibility, R commands can be bundled into a plain text file, called a "script" file, which can be executed using the source() command. The optional argument echo=TRUE for the source() command can be set to display each command and its output. The book Web site cited above includes

```
% R
R version 2.10.1 (2009-12-14)
Copyright (C) 2009 The R Foundation for Statistical Computing
ISBN 3-900051-07-0

R is free software and comes with ABSOLUTELY NO WARRANTY.
You are welcome to redistribute it under certain conditions.
Type 'license()' or 'licence()' for distribution details.

  Natural language support but running in an English locale

R is a collaborative project with many contributors.
Type 'contributors()' for more information and
'citation()' on how to cite R or R packages in publications.

Type 'demo()' for some demos, 'help()' for online help, or
'help.start()' for an HTML browser interface to help.
Type 'q()' to quit R.
```

```
 1   > 3+6
 2   [1] 9
 3   > 2*3
 4   [1] 6
 5   > x = c(4, 5, 3, 2)
 6   > x
 7   [1] 4 5 3 2
 8   > y = seq(1, 4)
 9   > y
10   [1] 1 2 3 4
11   > mean(x)
12   [1] 3.5
13   > sd(y)
14   [1] 1.290994
15   > ds = read.csv("http://www.math.smith.edu/r/data/help.csv")
16   > mean(ds$age)
17   [1] 35.65342
18   > mean(age)
19   Error in mean(age) : object "age" not found
20   > attach(ds)
21   > mean(age)
22   [1] 35.65342
23   > age[1:30]
24   [1] 37 37 26 39 32 47 49 28 50 39 34 58 53 58 60 36 28 35 29 27 27
25   [22] 41 33 34 31 39 48 34 32 35
26   > detach(ds)
27   > q()
28   Save workspace image? [y/n/c]: n
```

Figure 1.3: Sample session.

the R source code for the examples. The sample session in Figure 1.3 can be executed by running the following command.

```
> source("http://www.math.smith.edu/r/examples/sampsess.R",
    echo=TRUE)
```

The examples at the end of each chapter can be executed by running the following command, where X is replaced by the desired chapter number.

```
> source("http://www.math.smith.edu/r/examples/chapterX.R",
    echo=TRUE)
```

Many add-on packages need to be installed prior to running the examples (see 1.7.1). To facilitate this process, we have created a script file to load them in one step:

```
> source("http://www.math.smith.edu/r/examples/install.R",
    echo=TRUE)
```

If these libraries are not installed (1.7.1), running the example files at the end of the chapters will generate error messages.

1.2.2   Batch mode

In addition, R can be run in batch (noninteractive) mode from a command line interface:

```
% R CMD BATCH file.R
```

This will run the commands contained within `file.R` and put all output into `file.Rout`. If an additional argument is given to the command, the value of that argument is used as the filename containing the output.

Special note for Windows users: to use R in batch mode, you will need to include R.exe in your path. In Windows XP, this can be accomplished as follows, assuming the default installation directory set up for R version 2.10.1. For other versions of R or nondefault installations, the appropriate directory needs to be specified in the last step.

1. Right-click on "My Computer."

2. Click "Properties."

3. Select "Advanced" tab.

4. Press "Environment Variables" button.

5. Click "Path" (to highlight it).

6. Add `c:\program files\R\R-2.10.1\bin`

Once this is set up, the previously described R CMD BATCH syntax will work. Alternatively, changing the current directory to the one printed in step 6 will allow use of the BATCH syntax without these steps.

1.3 Using the R Commander graphical interface

R Commander [19] provides a sophisticated graphical user interface with menus, buttons, information fields and links to help files. Figure 1.4 displays the view after a linear regression model is fit. R Commander can be installed using the command install.packages("Rcmdr") (see also 1.7.1) and run using library(Rcmdr).

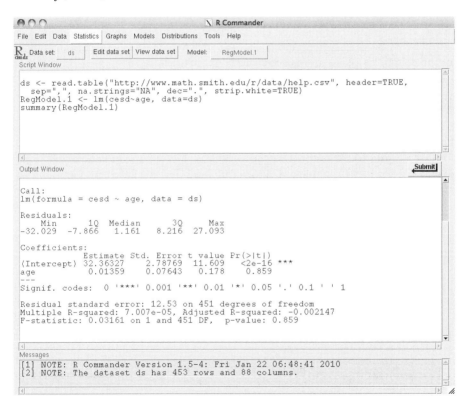

Figure 1.4: R Commander graphical user interface.

1.4   Learning R and getting help

An excellent starting point with R can be found in the *Introduction to R*, available from CRAN.

The system includes online documentation, though it can sometimes be challenging to comprehend. Each command has an associated help file that describes usage, lists arguments, provides details of actions, references, lists other related functions, and includes examples of its use. The help system is invoked using the command:

```
> ?function
```

or

```
> help(function)
```

where `function` is the name of the function of interest. As an example, the help file for the `mean()` function is accessed by the command `help(mean)`. The output from this command is provided in Figure 1.5.

The documentation describes the `mean()` function as a generic function for the (trimmed) arithmetic mean, with arguments `x` (an R object), `trim` (the fraction of observations to trim, default=0, trim=0.5 is equivalent to the median), and `na.rm` (should missing values be deleted, default is `na.rm=FALSE`). The function is described as returning a vector with the appropriate mean applied column by column. Related functions include `weighted.mean()` and `mean.POSIXct()`. Examples of many functions are available by running the `example()` function:

```
> example(mean)
mean> x <- c(0:10, 50)
mean> xm <- mean(x)
mean> c(xm, mean(x, trim = 0.10))
[1] 8.75 5.50

mean> mean(USArrests, trim = 0.2)
  Murder  Assault UrbanPop    Rape
    7.42   167.60    66.20   20.16
```

Other useful resources are `help.start()`, which provides a set of online manuals and `help.search()`, which can be used to look up entries by description. The `apropos()` command returns any functions in the current search list (including packages that have been loaded) that match a given pattern. This facilitates searching for a function based on what it does, as opposed to its name. The syntax `??pattern` can be used to search for strings in the documentation, while the `RSiteSearch()` function can be used to search for key words or phrases in the R-help mailing list archives. In addition, the `findFn()` function in `library(sos)` provides powerful search capabilities.

```
mean                    package:base                R Documentation

Arithmetic Mean

Description:
    Generic function for the (trimmed) arithmetic mean.

Usage:
    mean(x, ...)
    ## Default S3 method:
    mean(x, trim = 0, na.rm = FALSE, ...)

Arguments:
        x: An R object.  Currently there are methods for numeric data
           frames, numeric vectors and dates.  A complex vector is
           allowed for 'trim = 0', only.

     trim: the fraction (0 to 0.5) of observations to be trimmed from
           each end of 'x' before the mean is computed. Values of trim
           outside that range are taken as the nearest endpoint.

    na.rm: a logical value indicating whether 'NA' values should be
           stripped before the computation proceeds.

      ...: further arguments passed to or from other methods.

Value:
    For a data frame, a named vector with the appropriate method being
    applied column by column.

    If 'trim' is zero (the default), the arithmetic mean of the values
    in 'x' is computed, as a numeric or complex vector of length one.
    If 'x' is not logical (coerced to numeric), integer, numeric or
    complex, 'NA' is returned, with a warning.

    If 'trim' is non-zero, a symmetrically trimmed mean is computed
    with a fraction of 'trim' observations deleted from each end
    before the mean is computed.

References:
    Becker, R. A., Chambers, J. M. and Wilks, A. R. (1988) _The New S
    Language_. Wadsworth & Brooks/Cole.

See Also:
    `weighted.mean' `mean.POSIXct' `colMeans' for row and column means.

Examples:
    x <- c(0:10, 50)
    xm <- mean(x)
    c(xm, mean(x, trim = 0.10))
    mean(USArrests, trim = 0.2)
```

Figure 1.5: Documentation on the `mean()` function.

Other resources for help available from CRAN include the *Introduction to R* (described earlier) and the R-help mailing list (see also Section 1.8, support). New users are also encouraged to read the FAQ (frequently asked questions) list.

1.5   Fundamental structures: Objects, classes, and related concepts

Here we provide a brief introduction to data structures. The *Introduction to R* (discussed in Section 1.2) provides more comprehensive coverage.

1.5.1   Objects and vectors

Almost everything is an object, which may be initially disconcerting to a new user. An object is simply something that R can operate on. Common objects include vectors, matrices, arrays, factors (see 2.4.16), dataframes (akin to datasets in other packages), lists, and functions.

The basic variable structure is a vector. Vectors can be created using the = or <- assignment operators (which assigns the evaluated expression on the right-hand side of the operator to the object on the left-hand side). For instance, the following code creates a vector of length 6 using the c() function to concatenate scalars.

```
> x = c(5, 7, 9, 13, -4, 8)
```

Other assignment operators exist, as well as the `assign()` function (see Section 2.11.8 or `help("<-")` for more information).

1.5.2   Indexing

Since vector operations are so common, it is important to be able to access (or index) elements within these vectors. Many different ways of indexing vectors are available. Here, we introduce several of these, using the above example. The command x[2] would return the second element of x (the scalar 7), and x[c(2,4)] would return the vector (7,13). The expressions x[c(T,T,T,T,T,F)], x[1:5] (first through fifth element) and x[-6] (all elements except the sixth) would all return a vector consisting of the first five elements in x. Knowledge and basic comfort with these approaches to vector indexing is important to effective use of R.

Operations should be carried out wherever possible in a vector fashion (this is different from some other packages, where data manipulation operations are typically carried out an observation at a time). For example, the following commands demonstrate the use of comparison operators.

```
> rep(8, length(x))
[1] 8 8 8 8 8 8
> x>rep(8, length(x))
[1] FALSE FALSE  TRUE  TRUE FALSE FALSE
> x>8
[1] FALSE FALSE  TRUE  TRUE FALSE FALSE
```

Note that vectors are reused as needed, as in the last comparison. Only the third and fourth elements of x are greater than 8. The function returns a logical value of either TRUE or FALSE. A count of elements meeting the condition can be generated using the sum() function.

```
sum(x>8)
[1] 2
```

The code to create a vector of values greater than 8 is given below.

```
> largerthan8 = x[x>8]
> largerthan8
[1] 9 13
```

The command x[x>8] can be interpreted as "return the elements of x for which x is greater than 8." This construction is sometimes difficult for some new users, but is powerful and elegant. Examples of its application in the book can be found in Sections 2.4.18 and 2.13.4.

Other comparison operators include == (equal), >= (greater than or equal), <= (less than or equal and != (not equal). Care needs to be taken in the comparison using == if noninteger values are present (see 2.8.5). The which() function (see 3.1.1) can be used to find observations that match a given expression.

1.5.3 Operators

There are many operators defined to carry out a variety of tasks. Many of these were demonstrated in the sample section (assignment, arithmetic) and above examples (comparison). Arithmetic operations include +, -, *, /, ^ (exponentiation), %% (modulus), and &/& (integer division). More information about operators can be found using the help system (e.g., ?"+"). Background information on other operators and precedence rules can be found using help(Syntax).

R supports Boolean operations (OR, AND, NOT, and XOR) using the |, &, ! operators and the xor() function, respectively.

1.5.4 Lists

Lists are ordered collections of objects that are indexed using the [[operator or through named arguments.

```
> newlist = list(x1="hello", x2=42, x3=TRUE)
> is.list(newlist)
[1] TRUE
> newlist
$x1
[1] "hello"
$x2
[1] 42
$x3
[1] TRUE
> newlist[[2]]
[1] 42
> newlist$x2
[1] 42
```

1.5.5 Matrices

Matrices are rectangular objects with two dimensions. We can create a 2×3 matrix using our existing vector from Section 1.5.1, display it, and test for its type with the following commands.

```
> A = matrix(x, 2, 3)
> A
      [,1] [,2] [,3]
[1,]    5    9   -4
[2,]    7   13    8
> dim(A)
[1] 2 3
> # is A a matrix?
> is.matrix(A)
[1] TRUE
> is.vector(A)
[1] FALSE
> is.matrix(x)
[1] FALSE
```

Comments can be included: any input given after a # character until the next new line is ignored.

Indexing for matrices is done in a similar fashion as for vectors, albeit with a second dimension (denoted by a comma).

```
> A[2,3]
[1] 8
> A[,1]
[1] 5 7
> A[1,]
[1]  5  9 -4
```

1.5.6 Dataframes

The main way to access data is through a dataframe, which is more general than a matrix. This rectangular object, similar to a dataset in other statistics packages, can be thought of as a matrix with columns of vectors of different types (as opposed to a matrix, which consists of vectors of the same type). The functions data.frame(), read.csv(), (see Section 2.1.5) and read.table() (see 2.1.2) return dataframe objects. A simple dataframe can be created using the data.frame() command. Access to subelements is achieved using the $ operator as shown below (see also help(Extract)).

In addition, operations can be performed by column (e.g., calculation of sample statistics):

```
> y = rep(11, length(x))
> y
[1] 11 11 11 11 11 11
> ds = data.frame(x, y)
> ds
   x  y
1  5 11
2  7 11
3  9 11
4 13 11
5 -4 11
6  8 11
> is.data.frame(ds)
[1] TRUE
> ds$x[3]
[1] 9
> mean(ds)
       x         y
6.333333 11.000000
> sd(ds)
       x        y
5.715476 0.000000
```

Note that use of `data.frame()` differs from the use of `cbind()` (see 2.5.5), which yields a matrix object.

```
> y = rep(11, length(x))
> y
[1] 11 11 11 11 11 11
> newmat = cbind(x, y)
> newmat
      x  y
[1,]  5 11
[2,]  7 11
[3,]  9 11
[4,] 13 11
[5,] -4 11
[6,]  8 11
> is.data.frame(newmat)
[1] FALSE
> is.matrix(newmat)
[1] TRUE
```

Dataframes are created from matrices using `as.data.frame()`, while matrices can be constructed using `as.matrix()` or `cbind()`.

Dataframes can be attached using the `attach(ds)` command (see 2.3.1). After this command, individual columns can be referenced directly (i.e., x instead of ds$x). By default, the dataframe is second in the search path (after the local workspace and any previously loaded packages or dataframes). Users are cautioned that if there is a variable x in the local workspace, this will be referenced instead of ds$x, even if `attach(ds)` has been run. Name conflicts of this type are a common problem and care should be taken to avoid them.

The `search()` function lists attached packages and objects. To avoid cluttering the R workspace, the command `detach(ds)` should be used once the dataframe is no longer needed. The `with()` and `within()` commands (see 2.3.1 and 6.1.3) can also be used to simplify reference to an object within a dataframe without attaching.

Sometimes a package (Section 1.7.1) will define a function (Section 1.6) with the same name as an existing function. This is usually harmless, but to reverse it, detach the package using the syntax `detach("package:PKGNAME")`, where PKGNAME is the name of the package (see 5.7.6).

The names of all variables within a given dataset (or more generally for subobjects within an object) are provided by the `names()` command. The names of all objects defined within an R session can be generated using the `objects()` and `ls()` commands, which return a vector of character strings. Objects within the workspace can be removed using the `rm()` command. To remove all objects, (carefully) run the command `rm(list=ls())`.

The `print()` and `summary()` functions can be used to display brief or more extensive descriptions, respectively, of an object. Running `print(object)` at the command line is equivalent to just entering the name of the object, i.e., `object`.

1.5.7 Attributes and classes

Objects have a set of associated attributes (such as names of variables, dimensions, or classes) which can be displayed or sometimes changed. While a powerful concept, this can often be initially confusing. For example, we can find the dimension of the matrix defined in Section 1.5.5.

```
> attributes(A)
$dim
[1] 2 3
```

Other types of objects include lists (ordered objects that are not necessarily rectangular, Section 1.5.4), regression models (objects of class `lm`), and formulas (e.g., y ~ x1 + x2). Examples of the use of formulas can be found in Sections 3.4.1 and 4.1.1.

Many objects have an associated Class attribute, which cause that object to inherit (or take on) properties depending on the class. Many functions have special capabilities when operating on a particular class. For example, when `summary()` is applied to a `lm` object, the `summary.lm()` function is called, while `summary.aov()` is called when an `aov` object is given as argument. The `class()` function returns the classes to which an object belongs, while the `methods()` function displays all of the classes supported by a function (e.g., `methods(summary)`).

The `attributes()` command displays the attributes associated with an object, while the `typeof()` function provides information about the object (e.g., logical, integer, double, complex, character, and list).

1.5.8 Options

The `options()` function can be used to change various default behaviors, for example, the default number of digits to display in output can be specified using the command `options(digits=n)` where `n` is the preferred number (see 2.13.1). The command `help(options)` lists all of the other settable options.

1.6 Built-in and user-defined functions

1.6.1 Calling functions

Fundamental actions are carried out by calling functions (either built-in or user-defined), as seen previously. Multiple arguments may be given, separated by commas. The function carries out operations using these arguments using a series of predefined expressions, then returns values (an object such as a vector or list) that are displayed (by default) or saved by assignment to an object.

As an example, the `quantile()` function takes a vector and returns the minimum, 25th percentile, median, 75th percentile and maximum, though if an optional vector of quantiles is given, those are calculated instead.

```
> vals = rnorm(1000) # generate 1000 standard normals
> quantile(vals)
      0%      25%      50%      75%     100%
-3.1180  -0.6682   0.0180   0.6722   2.8629
> quantile(vals, c(.025, .975))
 2.5% 97.5%
-2.05   1.92
```

Return values can be saved for later use.

```
> res = quantile(vals, c(.025, .975))
> res[1]
 2.5%
-2.05
```

Options are available for many functions. These are named arguments for the function, and are generally added after the other arguments, also separated by commas. The documentation specifies the default action if named arguments (options) are not specified. For the `quantile()` function, there is a `type()` option which allows specification of one of nine algorithms for calculating quantiles. Setting `type=3` specifies the "nearest even order statistic" option, which is the default for some other packages.

```
res = quantile(vals, c(.025, .975), type=3)
```

Some functions allow a variable number of arguments. An example is the `paste()` function (see usage in 2.4.6). The calling sequence is described in the documentation in the following manner.

```
paste(..., sep=" ", collapse=NULL)
```

To override the default behavior of a space being added between elements output by `paste()`, the user can specify a different value for `sep` (see 7.1.2).

1.6.2 Writing functions

One of the strengths of R is its extensibility, which is facilitated by its programming interface. A new function (here named `newfun`) is defined in the following way.

```
newfun = function(arglist) body
```

The body is made up of a series of commands (or expressions), enclosed between an opening { and a closing }. Here, we demonstrate a function to calculate the estimated confidence interval for a mean from Section 3.1.7.

```
# calculate a t confidence interval for a mean
ci.calc = function(x, ci.conf=.95) {
    sampsize = length(x)
    tcrit = qt(1-((1-ci.conf)/2), sampsize)
    mymean = mean(x)
    mysd = sd(x)
    return(list(civals=c(mymean-tcrit*mysd/sqrt(sampsize),
            mymean+tcrit*mysd/sqrt(sampsize)),
            ci.conf=ci.conf))
}
```

Here the appropriate quantile of the t distribution is calculated using the `qt()` function, and the appropriate confidence interval is calculated and returned as a list. The function is stored in the object `ci.calc`, which can then be run interactively on our vector from Section 1.5.1.

```
> ci.calc(x)
$civals
[1]   0.6238723 12.0427943
$ci.conf
[1] 0.95
```

If only the lower confidence interval is needed, this can be saved as an object.

```
> lci = ci.calc(x)$civals[1]
> lci
[1] 0.6238723
```

The default confidence level is 95%; this can be changed by specifying a different value as the second argument.

```
> ci.calc(x, ci.conf=0.90)
$civals
[1]   1.799246 10.867421

$ci.conf
[1] 0.9
```

This is equivalent to running `ci.calc(x, 0.90)`. Other sample programs can be found in Sections 2.4.22 and 3.6.4 as well as Chapter 7.

1.6.3 The `apply` family of functions

Operations are most efficiently carried out using vector or list operations rather than looping. The `apply()` function can be used to perform many actions. While somewhat subtle, the power of the vector language can be seen in this example. The `apply()` command is used to calculate column means or row means of the previously defined matrix in one fell swoop.

```
> A
     [,1] [,2] [,3]
[1,]    5    9   -4
[2,]    7   13    8
> apply(A, 2, mean)
[1]  6 11  2
> apply(A, 1, mean)
[1] 3.333333 9.333333
```

Option 2 specifies that the mean should be calculated for each column, while option 1 calculates the mean of each row. Here we see some of the flexibility of the system, as functions (such as `mean()`) are also objects that can be passed as arguments to functions.

Other related functions include `lapply()`, which is helpful in avoiding loops when using lists, `sapply()` (see 2.3.2), and `mapply()` to do the same for dataframes and matrices, respectively, and `tapply()` (see 3.1.2) to perform an action on subsets of an object.

1.7   Add-ons: Libraries and packages

1.7.1   Introduction to libraries and packages

Additional functionality is added through packages, which consist of libraries of bundled functions, datasets, examples and help files that can be downloaded from CRAN. The function `install.packages()` or the windowing interface under *Packages and Data* must be used to download and install packages. The `library()` function can be used to load a previously installed package (that has been previously made available through use of the `install.packages()` function). As an example, to install and load the `Hmisc` package, two commands are needed:

```
install.packages("Hmisc")
library(Hmisc)
```

Once a package has been installed, it can be loaded whenever a new session is run by executing the function `library(libraryname)`. A package only needs to be installed once for a given version of R.

If a package is not installed, running the `library()` command will yield an error. Here we try to load the Zelig package (which had not yet been installed).

```
> library(Zelig)
Error in library(Zelig) : there is no package called 'Zelig'
```

```
> install.packages("Zelig")
trying URL 'http://cran.stat.auckland.ac.nz/cran/bin/macosx/
    leopard/contrib/2.10/Zelig_3.4-5.tgz'
Content type 'application/x-gzip' length 14460464 bytes (13.8 Mb)
opened URL
==================================================
downloaded 13.8 Mb


The downloaded packages are in
/var/folders/Tmp/RtmpXaN7Kk/downloaded_packages

> library(Zelig)
Loading required package: MASS
Loading required package: boot
##  Zelig (Version 3.4-5, built: 2009-03-13)
##  Please refer to http://gking.harvard.edu/zelig for full
##  documentation or help.zelig() for help with commands and
##  models supported by Zelig.
##
##  To cite individual Zelig models, please use the citation
##  format printed with each model run and in the documentation.
```

A user can test whether a package is loaded by running require(packagename); this will load the library if it is installed, and generate an error message if it is not. The update.packages() function should be run periodically to ensure that packages are up-to-date.

As of February 2010, there were 2172 packages available from CRAN [20]. While each of these has met a minimal standard for inclusion, it is important to keep in mind that packages are created by individuals or small groups, and not endorsed by the R core group. As a result, they do not necessarily undergo the same level of testing and quality assurance that the core R system does. Hadley Wickham's Crantastic (http://crantastic.org) is a community site that reviews and tags CRAN packages.

1.7.2   CRAN task views

A very useful resource for finding packages are the *Task Views* on CRAN (http://cran.r-project.org/web/views). These are listings of relevant packages within a particular application area (such as multivariate statistics, psychometrics, or survival analysis). Table 1.1 displays the Task Views available as of January 2010.

Bayesian	Bayesian inference
ChemPhys	Chemometrics and computational physics
Clinical Trials	Design, monitoring, and analysis of clinical trials
Cluster	Cluster analysis & finite mixture models
Distributions	Probability distributions
Econometrics	Computational econometrics
Environmetrics	Analysis of ecological and environmental data
Experimental Design	Design and analysis of experiments
Finance	Empirical finance
Genetics	Statistical genetics
Graphics	Graphic displays, dynamic graphics, graphic devices, and visualization
gR	Graphical models in R
HPC	High-performance and parallel computing with R
Machine Learning	Machine and statistical learning
Medical Imaging	Medical image analysis
Multivariate	Multivariate statistics
NLP	Natural language processing
Optimization	Optimization and mathematical programming
Pharmacokinetics	Analysis of pharmacokinetic data
Psychometrics	Psychometric models and methods
Robust	Robust statistical methods
Social Sciences	Statistics for the social sciences
Spatial	Analysis of spatial data
Survival	Survival analysis
Time Series	Time series analysis

Table 1.1: CRAN Task Views

1.7.3   Installed libraries and packages

Running the command `library(help="libraryname"))` will display information about an installed package (assuming that it has been installed). Entries in the book that utilize packages include a line specifying how to access that library (e.g., `library(foreign)`). Vignettes showcase the ways that a package can be useful in practice: the command `vignette()` will list installed vignettes.

As of January 2010, the R distribution comes with the following packages.

base Base R functions

datasets Base R datasets

grDevices Graphics devices for base and grid graphics

graphics R functions for base graphics

grid A rewrite of the graphics layout capabilities

methods Formally defined methods and classes, plus programming tools

splines Regression spline functions and classes

stats R statistical functions

stats4 Statistical functions using S4 classes

tcltk Interface and language bindings to Tcl/Tk GUI elements

tools Tools for package development and administration

utils R utility functions

These are available without having to run the `library()` command and are effectively part of the base system.

1.7.4 Recommended packages

A set of recommended packages are to be included in all binary distributions of R. As of January 2010, these included the following list.

KernSmooth Functions for kernel smoothing (and density estimation)

MASS Functions and datasets from the main package of Venables and Ripley, "Modern Applied Statistics with S"

Matrix A Matrix package

boot Functions and datasets for bootstrapping

class Functions for classification (k-nearest neighbor and LVQ)

cluster Functions for cluster analysis

codetools Code analysis tools

foreign Functions for reading and writing data stored by statistical software like Minitab, S, SAS, SPSS, Stata, Systat, etc.

lattice Lattice graphics, an implementation of Trellis Graphics functions

mgcv Routines for GAMs and other generalized ridge regression problems

nlme Fit and compare Gaussian linear and nonlinear mixed-effects models

nnet Software for single hidden layer perceptrons ("feed-forward neural networks") and for multinomial log-linear models

rpart Recursive partitioning and regression trees

spatial Functions for kriging and point pattern analysis from MASS

survival Functions for survival analysis, including penalized likelihood

1.7.5   Packages referenced in the book

Other packages referenced in the book but not included in the R distribution are listed below (to see more information about a particular package, run the command `citation(package="packagename")`.

amer Additive mixed models with lme4

chron Chronological objects which can handle dates and times

circular Circular statistics

coda Output analysis and diagnostics for MCMC

coin Conditional inference procedures in a permutation test framework

dispmod Dispersion models

ellipse Functions for drawing ellipses and ellipse-like confidence regions

elrm Exact logistic regression via MCMC

epitools Epidemiology tools

exactRankTests Exact distributions for rank and permutation tests

frailtypack Frailty models using maximum penalized likelihood estimation

gam Generalized additive models

gee Generalized estimation equation solver

GenKern Functions for kernel density estimates

ggplot2 An implementation of the Grammar of Graphics

gmodels Various R programming tools for model fitting

gtools Various R programming tools

Hmisc Harrell miscellaneous functions

irr Various coefficients of interrater reliability and agreement

lars Least angle regression, lasso and forward stagewise

lme4 Linear mixed-effects models using S4 classes

lmtest Testing linear regression models

lpSolve Interface to Lp_solve v. 5.5 to solve linear/integer programs

maps Draw geographical maps

Matching Propensity score matching with balance optimization

MCMCpack Markov Chain Monte Carlo (MCMC) package

mice Multivariate imputation by chained equations

mitools Tools for multiple imputation of missing data

mix Multiple imputation for mixed categorical and continuous data

multcomp Simultaneous inference in general parametric models

multilevel Multilevel functions

nnet Feed-forward neural networks and multinomial log-linear models

nortest Tests for normality

odfWeave Sweave processing of Open Document Format (ODF) files

plotrix Various plotting functions

plyr Tools for splitting, applying and combining data

prettyR Pretty descriptive stats

pscl Political science computational laboratory, Stanford University

pwr Basic functions for power analysis

quantreg Quantile regression

Rcmdr R Commander

RColorBrewer ColorBrewer palettes

reshape Flexibly reshape data

rms Regression modeling strategies

RMySQL R interface to the MySQL database

ROCR Visualizing the performance of scoring classifiers

RSQLite SQLite interface for R

scatterplot3d 3D scatterplot

sos Search help pages of R packages

sqldf Perform SQL selects on R data frames

survey Analysis of complex survey samples

tmvtnorm Truncated multivariate normal distribution

vcd Visualizing categorical data

VGAM Vector generalized linear and additive models

XML Tools for parsing and generating XML within R

Zelig Everyone's statistical software [31]

Many of these must be installed and loaded prior to use (see `install.packages()`, `require()`, `library()` and Section 1.7.1). To facilitate this process, we have created a script file to load those needed to replicate the example code in one step (see 1.2.1).

1.7.6   Datasets available with R

A number of datasets are available within the `datasets` package that is included in the R distribution. The `data()` function lists these, while the `package` option can be used to specify datasets from within a specific package.

1.8   Support and bugs

Since R is a free software project written by volunteers, there are no paid support options available directly from the R Foundation. However, extensive resources are available to help users.

In addition to the manuals, FAQs, newsletter, wiki, task views, and books listed on the `www.r-project.org` Web page, there are a number of mailing lists that exist to help answer questions. Because of the volume of postings, it is important to carefully read the posting guide at `http://www.r-project.org/posting-guide.html` prior to submitting a question. These guidelines are intended to help leverage the value of the list, to avoid embarrassment, and to optimize the allocation of limited resources to technical issues.

As in any general purpose statistical software package, bugs exist. More information about the process of determining whether and how to report a problem can be found in the R FAQ as well as via the information available using `help(bug.report)`.

Chapter 2

Data management

This chapter reviews basic data management, beginning with accessing external datasets, such as those stored in spreadsheets, ASCII files, or foreign formats. Important tasks such as creating datasets and manipulating variables are discussed in detail. In addition, key mathematical, statistical, and probability functions are introduced.

2.1 Input

In this section we address aspects of data input. Data are organized in dataframes (1.5.6), or connected series of rectangular arrays, which can be saved as platform independent objects.

2.1.1 Native dataset

Example: See 5.7

```
load(file="dir_location\\savedfile")   # Windows only
load(file="dir_location/savedfile")    # other OS
```

Forward slash is supported as a directory delimiter on all operating systems; a double backslash is also supported under Windows. The file savedfile is created by save() (see 2.2.1).

2.1.2   Fixed format text files

See also 2.1.4 (read more complex fixed files) and 7.4.1 (read variable format files)

```
# Windows only
ds = read.table("dir_location\\file.txt", header=TRUE)


# all OS (including Windows)
ds = read.table("dir_location/file.txt", header=TRUE)
```

Forward slash is supported as a directory delimiter on all operating systems; a double backslash is also supported under Windows. If the first row of the file includes the name of the variables, these entries will be used to create appropriate names (reserved characters such as '$' or '[' are changed to '.') for each of the columns in the dataset. If the first row does not include the names, the `header` option can be left off (or set to `FALSE`), and the variables will be called `V1`, `V2`, ... `Vn`. The `read.table()` function can support reading from a URL as a filename (see also 2.1.7). Files can be browsed interactively using `file.choose()` (see 2.7.7).

2.1.3   Other fixed files

See also 2.1.4 (read more complex fixed files) and 7.4.1 (read variable format files)

```
ds = readLines("file.txt")
```

or

```
ds = scan("file.txt")
```

The `readLines()` function returns a character vector with length equal to the number of lines read (see also `file()`). A limit on the number of lines to be read can be specified through the `nrows` option. The `scan()` function returns a vector.

2.1.4   Reading more complex text files

See also 2.1.2 (read fixed files) and 7.4.1 (read variable format files).

Text data files often contain data in special formats. One common example is date variables. As an example below we consider the following data.

```
1 AGKE 08/03/1999 $10.49
2 SBKE 12/18/2002 $11.00
3 SEKK 10/23/1995 $5.00
```

```
tmpds = read.table("file_location/filename.dat")
id = tmpds$V1
initials = tmpds$V2
datevar = as.Date(as.character(tmpds$V3), "%m/%d/%Y")
cost = as.numeric(substring(tmpds$V4, 2))
ds = data.frame(id, initials, datevar, cost)
rm(tmpds, id, initials, datevar, cost)
```

This task is accomplished by first reading the dataset (with default names from `read.table()` denoted `V1` through `V4`). These objects can be manipulated using `as.character()` to undo the default coding as factor variables, and coerced to the appropriate data types. For the `cost` variable, the dollar signs are removed using the `substring()` function (Section 2.4.4). Finally, the individual variables are gathered together as a dataframe.

2.1.5   Comma-separated value (CSV) files

Example: See 2.13.1

Comma-separated value (CSV) files are a common data interchange format that are straightforward to read and write.

```
ds = read.csv("dir_location/file.csv")
```

A limit on the number of lines to be read can be specified through the `nrows` option. The command `read.csv(file.choose())` can be used to browse files interactively (see Section 2.7.7). The comma-separated file can be given as a URL (see 2.1.7).

2.1.6   Reading datasets in other formats

Example: See 6.6

```
library(foreign)
ds = read.dbf("filename.dbf")          # DBase
ds = read.epiinfo("filename.epiinfo")  # Epi Info
ds = read.mtp("filename.mtp")          # Minitab worksheet
ds = read.octave("filename.octave")    # Octave
ds = read.ssd("filename.ssd")          # SAS version 6
ds = read.xport("filename.xport")      # SAS XPORT file
ds = read.spss("filename.sav")         # SPSS
ds = read.dta("filename.dta")          # Stata
ds = read.systat("filename.sys")       # Systat
```

The `foreign` library can read Stata, Epi Info, Minitab, Octave, SAS version 6, SAS Xport, SPSS, and Systat files (with the caveat that SAS version 6 files may be platform dependent). The `read.ssd()` function will only work if SAS is installed on the local machine (as it needs to run SAS in order to read the dataset).

2.1.7   URL

Example: See 3.6.1

Data can be read from the Web by specifying a uniform resource locator (URL). Many of the data input functions also support accessing data in this manner (see also 2.1.3).

```
urlhandle = url("http://www.math.smith.edu/r/testdata")
ds = readLines(urlhandle)
```

or

```
ds = read.table("http://www.math.smith.edu/r/testdata")
```

or

```
ds = read.csv("http://www.math.smith.edu/r/file.csv")
```

The `readLines()` function reads arbitrary text, while `read.table()` can be used to read a file with cases corresponding to lines and variables to fields in the file (the `header` option sets variable names to entries in the first line). The `read.csv()` function can be used to read comma-separated values. Access through proxy servers as well as specification of username and passwords is provided by the function `download.file()`. A limit on the number of lines to be read can be specified through the `nrows` option.

2.1.8   XML (extensible markup language)

A sample (flat) XML form of the HELP dataset can be found at `http://www.math.smith.edu/r/data/help.xml`. The first 10 lines of the file consist of:

```
<?xml version="1.0" encoding="iso-8859-1" ?>
<TABLE>
   <HELP>
      <id> 1 </id>
      <e2b1 Missing="." />
      <g1b1> 0 </g1b1>
      <i11 Missing="." />
      <pcs1> 54.2258263 </pcs1>
      <mcs1> 52.2347984 </mcs1>
      <cesd1> 7 </cesd1>
```

Here we consider reading simple files of this form. While support is available for reading more complex types of XML files, these typically require considerable additional sophistication.

```
library(XML)
urlstring = "http://www.math.smith.edu/r/data/help.xml"
doc = xmlRoot(xmlTreeParse(urlstring))
tmp = xmlSApply(doc, function(x) xmlSApply(x, xmlValue))
ds = t(tmp)[,-1]
```

The XML library provides support for reading XML files. The xmlRoot() function opens a connection to the file, while xmlSApply() and xmlValue() are called recursively to process the file. The returned object is a character matrix with columns corresponding to observations and rows corresponding to variables, which in this example are then transposed.

2.1.9 Data entry

```
x = numeric(10)
data.entry(x)
```
or
```
x1 = c(1, 1, 1.4, 123)
x2 = c(2, 3, 2, 4.5)
```

The data.entry() function opens a spreadsheet that can be used to edit or otherwise change a vector or dataframe. In this example, an empty numeric vector of length 10 is created to be populated. The data.entry() function differs from the edit() function, which leaves the objects given as argument unchanged, returning a new object with the desired edits (see also fix()).

2.2 Output

2.2.1 Save a native dataset

Example: See 2.13.3

```
save(robject, file="savedfile")
```

An object (typically a dataframe, or a list of objects) can be read back into R using load() (see 2.1.1).

2.2.2 Creating files for use by other packages

See also 2.2.5 (write XML)   *Example:* See 2.13.3

```
library(foreign)
write.dta(ds, "filename.dta")
write.dbf(ds, "filename.dbf")
write.foreign(ds, "filename.dat", "filename.sas", package="SAS")
```

Support for writing datasets for use in other software packages is provided in the `foreign` library. It is possible to write files directly in Stata format (see `write.dta()`) or DBF format (see `write.dbf()` or create files with fixed fields as well as the code to read the file from within Stata, SAS, or SPSS using `write.foreign()`).

As an example with a dataset with two numeric variables X_1 and X_2, the call to `write.foreign()` creates one file with the data and the SAS command file `filename.sas`, with the following contents.

```
data ds;
   infile "file.dat" dsd lrecl=79;
   input x1 x2;
run;
```

Similar code is created using SPSS syntax by calling `write.foreign()` with appropriate `package` option. Both functions require access to SAS or SPSS to complete the transfer.

2.2.3   Creating datasets in text format

Example: See 2.13.3

```
write.csv(ds, file="full_file_location_and_name")
```

or

```
write.table(ds, file="full_file_location_and_name")
```

The `sep` option to `write.table()` can be used to change the default delimiter (space) to an arbitrary value.

2.2.4   Creating HTML formatted output

Output can be created using HTML format to facilitate display on the Web.

```
library(prettyR)
htmlize("script.R", title="mytitle", echo=TRUE)
```

The `htmlize()` function within `library(prettyR)` can be used to produce HTML (hypertext markup language) from a script file (see 1.2.1). The `cat()` function is used inside the script file (here denoted by `script.R`) to generate output. The `hwriter` library also supports writing objects in HTML format.

2.2.5   Creating XML datasets and output

The `XML` library provides support for writing XML files (see also 2.1.6, read foreign files, and further resources).

2.2.6   Displaying objects

Example: See 2.13.4

```
print(ds)
```

or

```
View(ds)
```

or

```
newds = edit(ds)
```

or

```
ds[1:10,]
ds[,2:3]
```

or

```
ds[,c("x1", "x3", "xk", "x2"]
```

The `print()` function lists the contents of the dataframe (or any other object), while the `View()` function opens a navigable window with a read-only view. The contents can be changed using the `edit()` or `fix()` functions. Alternatively, any subset of the dataframe can be displayed on the screen using indexing, as in the final example. In the second to last example, the first 10 records (rows) are displayed, then all values for the second and third variables (columns) are printed on the console. Variables can also be specified by name using a character vector index (see 1.5.2), as in the last example. The `head()` function can be used to display the first (or last) values of a vector, dataset, or other object.

2.2.7   Displaying formatted output

Example: See 4.7.3

We demonstrate displaying formatted output by displaying numbers as U.S. currency (see also 2.2.6, values of variables in a dataset).

```
dollarcents = function(x)
  return(paste("$", format(round(x*100, 0)/100, nsmall=2),
    sep=""))
data.frame(x1, dollarcents(x3), xk, x2)
```

A function can be defined to format a vector as U.S. dollar and cents by using the `round()` function (see 2.8.4) to control the number of digits (2) to the right of the decimal.

The `cat()` function can be used to concatenate values and display them on the console (or route them to a file using the `file` option). More control on the appearance of printed values is available through use of `format()` (control of digits and justification), `sprintf()` (use of C-style string formatting) and `prettyNum()` (another routine to format using C-style specifications). The `head()` function will display the first values of an object. The `sink()` function can be used to redirect output to a file.

2.2.8 Number of digits to display

Example: See 2.13.1

```
options(digits=n)
```

The `options(digits=n)` command can be used to change the default number of decimal places to display in subsequent output (see also 1.5.8). To affect the actual significant digits in the data, use the `round()` function (see 2.8.4).

2.2.9 Automating reproducible analysis and output

It is straightforward to automate report generation using versions of literate programming (in the sense of Knuth) that facilitate "reproducible analysis" [37] The formatting of the HELP sections in this book were generated using a variant of this system due to Russell Lenth [38].

The `Sweave()` function combines documentation and R code in a source file, runs the code, taking the output (including text and graphics) and combining it into an intermediate file (e.g., LaTeX file). The `Stangle()` function just creates a file containing the code chunks which could be processed using `source()`. Other systems (e.g., `library(odfWeave)` and `StatWeave`) support Open Document Format (ODF) files.

2.3 Structure and meta-data

2.3.1 Access variables from a dataset

Example: See 2.13.1

Variable references must contain the name of the object which includes the variable, unless the object is `attach`ed (see below).

```
attach(ds)
detach(ds)
with(ds, mean(x))
```

The command `attach()` will make the variables within the named dataframe available in the workspace (otherwise they need to be accessed using the syntax `ds$var1`). The `detach()` function removes them from the workspace (and is recommended when the local version is no longer needed, to avoid name conflicts). The `with()` and `within()` functions provide another way to access variables within a dataframe without having to worry about later detaching the dataframe. Many functions (e.g., `lm()`) allow specification of a dataset to be accessed using the `data` option.

The `detach()` function is also used to remove a package from the workspace: more information can be found in Section 1.5.6. This is sometimes needed if a package overrides a built-in function. The command to detach a package that loaded using `library(packagename)` is `detach("package:packagename")`.

2.3.2 Names of variables and their types

Example: See 2.13.1

```
str(ds)
```

The command `sapply(ds, class)` will return the names and classes (e.g., numeric, integer or character) of each variable within a dataframe (see also 1.6.3). Running `summary(ds)` will provide an overview of the distribution of each column of the dataframe given as argument.

2.3.3 Rename variables in a dataset

```
names(ds)[names(ds)=="old1"] = "new1"
names(ds)[names(ds)=="old2"] = "new2"
```

or

```
ds = within(ds, {new1 = old1; new2 = old2; rm(old1, old2)})
```

or

```
library(reshape)
ds = rename(ds, c("old1"="new1", "old2"="new2"))
```

The `names()` function returns the list of names associated with an object (see 1.5.6). It is an efficient way to undertake this task, as it involves no copying of data (just a remapping of the names). The `edit()` function can be used to edit names and values.

2.3.4 Add comment to a dataset or variable

Example: See 2.13.1

To help facilitate proper documentation of datasets, it can be useful to provide some annotation or description.

```
comment(ds) = "This is a comment about the dataset"
```

The `attributes()` function (see 1.5.7) can be used to list all attributes, including any `comment()`, while the `comment()` function without an assignment will display the comment, if present.

2.4 Derived variables and data manipulation

This section describes the creation of new variables as a function of existing variables in a dataset.

2.4.1 Create string variables from numeric variables

```
stringx = as.character(numericx)
typeof(stringx)
typeof(numericx)
```

The `typeof()` function can be used to verify the type of an object (see also `class()`; possible values include `logical`, `integer`, `double`, `complex`, `character`, `raw`, `list`, `NULL`, `closure` (function), `special` and `builtin` (see also Section 1.5.7).

2.4.2 Create numeric variables from string variables

```
numericx = as.numeric(stringx)
typeof(stringx)
typeof(numericx)
```

The `typeof()` function can be used to verify the type of an object (see 2.4.1 and 1.5.7).

2.4.3 Concatenate vectors

```
newvector = c(x1, x2)
```

The `c()` concatenates a set of two (or more vectors), and returns a vector. Related vector functions for set-like operations include `union()`, `setdiff()`, `setequal()`, `intersect()`, `unique()`, `duplicated()`, `match()`, and the `%in%` operator.

2.4.4 Extract characters from string variables

```
get2through4 = substr(x, 2, 4)
get2throughend = substring(x, 2)
```

The arguments to `substr()` specify the input vector, start character position and end character position. For `substring()`, omitting the end character value takes all characters from the start to the end of the string.

2.4.5 Length of string variables

```
len = nchar(stringx)
```

The `nchar()` function returns a vector of lengths of each of the elements of the string vector given as argument, as opposed to the `length()` function (Section 2.4.19) returns the number of elements in a vector.

2.4.6 Concatenate string variables

```
newcharvar = paste(x1, " VAR2 ", x2, sep="")
```

The above code creates a character variable `newcharvar` containing the character vector X_1 (which may be coerced from a numeric object) followed by the string " VAR2 " then the character vector X_2. The `sep=""` option leaves no additional separation character between these three strings (the default is a single space, other characters may instead be specified).

2.4.7 Find strings within string variables

```
matches = grep("pat", stringx)
positions = regexpr("pat", stringx)
```

```
> x = c("abc", "def", "abcdef", "defabc")
> grep("abc", x)
[1] 1 3 4
> regexpr("abc", x)
[1]   1 -1   1   4
attr(,"match.length")
[1]   3 -1   3   3
> regexpr("abc", x) < 0
[1] FALSE  TRUE FALSE FALSE
```

The function `grep()` returns a list of elements in the vector given by `stringx` that match the given pattern, while the `regexpr()` function returns a numeric list of starting points in each string in the list (with -1 if there was no match). Testing `positions < 0` generates a vector of binary indicator of matches (TRUE=no match, FALSE=a match). The regular expressions supported within `grep` and other related routines are quite powerful. For an example, Boolean `OR` expressions can be specified using the | operator, while ! is the not operator. A comprehensive description of these can be found using `help(regex)` (see also 1.5.3).

2.4.8   Find approximate strings

```
> agrep("favor", "I ask a favour")
[1] 1
```

The `agrep()` function utilizes the Levenshtein edit distance (total number of insertions, deletions and substitutions required to transform one string into another). By default the threshold is 10% of the pattern length. The function returns the number of matches (see also 2.4.7).

2.4.9   Replace strings within string variables

Example: See 7.4.2

```
newstring = gsub("oldpat", "newpat", oldstring)
```

or

```
x = "oldpat123"
substr(x, 1, 6) = "newpat"
```

2.4.10   Split string into multiple strings

```
> x = "this is a test"
> split = " "
> strsplit(x, split)
[[1]]
[1] "this" "is"   "a"     "test"
> strsplit(x, "")
[[1]]
 [1] "t" "h" "i" "s" " " "i" "s" " " "a" " " "t" "e" "s" "t"
```

The function `strsplit()` returns a list of vectors split using the specified argument. If `split` is the null string, then the function returns a list of vectors of single characters.

2.4.11   Remove spaces around string variables

```
noleadortrail = sub(' +$', '', sub('^ +', '', stringx))
```

The arguments to `sub()` consist of a regular expression, a substitution value and a vector. In the first step, leading spaces are removed (nothing is included between single quotes), then a separate call to `sub()` is used to remove trailing spaces (in both cases replacing the spaces with the null string). If instead of spaces all trailing whitespaces (e.g., tabs, space characters) should be removed, the regular expression ' +$' should be replaced by '[[:space:]]+$'.

2.4.12 Upper to lower case

```
lowercasex = tolower(x)
```
or
```
lowercasex = chartr("ABCDEFGHIJKLMNOPQRSTUVWXYZ",
    "abcdefghijklmnopqrstuvwxzy", x)
```

The `toupper()` function can be used to convert to upper case. Arbitrary translations from sets of characters can be made using the `chartr()` function.

2.4.13 Create categorical variables from continuous variables

Example: See 2.13.5 and 5.7.7
```
newcat1 = (x >= minval) + (x >= cutpoint1) + ... +
    (x >= cutpointn)
```

Each expression within parentheses is a logical test returning 1 if the expression is true, 0 if not true, and NA if `x` is missing. More information about missing value coding can be found in Section 2.4.18. The `cut()` function (2.4.14) can also be used to divide a continuous variable into intervals.

2.4.14 Recode a categorical variable

A categorical variable can be recoded to have fewer levels.
```
tmpcat = oldcat
tmpcat[oldcat==val1] = newval1
tmpcat[oldcat==val2] = newval1
...
tmpcat[oldcat==valn] = newvaln
newcat = as.factor(tmpcat)
```
or
```
newcat = cut(x, breaks=c(val2, ..., valn),
    labels=c("Cut1", "Cut2", ..., "Cutn"), right=FALSE)
```

Creating the variable can be undertaken in multiple steps. A copy of the old variable is first made, then multiple assignments are made for each of the new levels, for observations matching the condition inside the index (see Section 1.5.2). In the final step, the categorical variable is coerced into a factor (class) variable (see also `help("%in%")`). Alternatively, the `cut()` function can be used to create the factor vector in one operation, by specifying the cut-scores and the labels.

2.4.15   Create a categorical variable using logic

See also 2.4.18 (missing values)   *Example:* See 2.13.5

Here we create a trichotomous variable `newvar` which takes on a missing value if the continuous non-negative variable `oldvar` is less than 0, 0 if the continuous variable is 0, value 1 for subjects in group A with values greater than 0 but less than 50 and for subjects in group B with values greater than 0 but less than 60, or value 2 with values above those thresholds.

```
tmpvar = rep(NA, length(oldvar))
tmpvar[oldvar==0] = 0
tmpvar[oldvar>0 & oldvar<50 & group=="A"] = 1
tmpvar[oldvar>0 & oldvar<60 & group=="B"] = 1
tmpvar[oldvar>=50 & group=="A"] = 2
tmpvar[oldvar>=60 & group=="B"] = 2
newvar = as.factor(tmpvar)
```

Creating the variable is undertaken in multiple steps in this example. A vector of the correct length is first created containing missing values. Values are updated if they match the conditions inside the vector index (see Section 1.5.2). Care needs to be taken in the comparison of `oldvar==0` if noninteger values are present (see 2.8.5).

2.4.16   Formatting values of variables

Example: See 4.7.3

Sometimes it is useful to display category names that are more descriptive than variable names. In general, we do not recommend using this feature (except potentially for graphical output), as it tends to complicate communication between data analysts and other readers of output. In this example, character labels are associated with a numeric variable (0=Control, 1=Low Dose, and 2=High Dose).

```
x = c(id1=0, id2=0, id3=1, id4=1, id5=2)
x = factor(x, 0:2, labels=c("Control", "Low Dose", "High Dose"))
```

For this example, the command `x` (equivalent to `print(x)` returns the following output.

```
> x
     id1       id2       id3       id4       id5
 Control   Control  Low Dose  Low Dose High Dose
Levels: Control Low Dose High Dose
```

Additionally, the `names()` function can be used to associate a variable with the identifier (which is by default the observation number). As an example, this can be used to display the name of a region with the value taken by a particular variable measured in that region.

2.4.17 Label variables

As with the values of the categories, sometimes it is desirable to have a longer, more descriptive variable name (see also formatting variables, 2.4.16).

```
comment(x) = "This is the label for the variable 'x'"
```

The label for the variable can be extracted using `comment(x)` with no assignment or via `attribute(x)$comment`. In addition, certain commands (e.g., `c()` and `data.frame()`) can give variables named labels.

```
> c(a=1, b=2)
a b
1 2
> data.frame(x1=1:2, x2=3:4)
  x1 x2
1  1  3
2  2  4
```

2.4.18 Account for missing values

Example: See 2.13.5

Missing values are ubiquitous in most real-world investigations. They are denoted by `NA`. This is a logical constant of length 1 which has no numeric equivalent. The missing value code is distinct from the character string value `"NA"`. The default behavior for most functions is to return `NA` if any of the input vectors have any missing values.

```
> mean(c(1, 2, NA))
[1] NA
> mean(c(1, 2, NA), na.rm=TRUE)
[1] 1.5
> sum(na.omit(c(1, 2, NA)))
[1] 3
> x = c(1, 3, NA)
> sum(!is.na(x))
[1] 2
> mean(x)
[1] NA
> mean(x, na.rm=TRUE)
[1] 2
```

The `na.rm` option is used to override the default behavior and omit missing values and calculate the result on the complete cases (this or related options are available for many functions). The `!` (not) operator allows counting of the number of observed values (since `is.na()` returns a logical set to TRUE if an observation is missing). Values can be recoded to missing, as well as omitted (see 1.4).

```
# remap values of x with missing value code of 999 to missing
x[x==999] = NA
```
or
```
# set 999's to missing
is.na(x) = x==999
# returns a vector of logicals
is.na(x)
# removes observations (rows) that are missing on that variable
na.omit(x)
# removes observations (rows) that are missing any variable
na.omit(ds)

library(Hmisc)
# display patterns of missing variables in a dataframe
na.pattern(ds)
```

The default of returning NA for functions operating on objects with missing values can be overridden using options for a particular function by using na.omit(), adding the na.rm=TRUE option (e.g., for the mean() function) or specifying an na.action() (e.g., for the lm() function). Common na.action() functions include na.exclude(), na.omit(), and na.fail(). Arbitrary numeric missing values (999 in this example) can be mapped to R missing value codes using indexing and assignment. Here all values of x that are 999 are replaced by the missing value code of NA. The is.na() function returns a logical vector with TRUE corresponding to missing values (code NA in R). Input functions like scan() and read.table() have the default argument na.strings="NA". This can be used to recode on input for situations where a numeric missing value code has been used. R has other kinds of "missing" values, corresponding to floating point standards (see also is.infinite() and is.nan()).

The na.pattern() function can be used to determine the different patterns of missing values in a dataset. The na.omit() function returns the dataframe with missing values omitted (if a value is missing for a given row, all observations are removed, aka listwise deletion). More sophisticated approaches to handling missing data are generally recommended (see 7.6).

2.4.19 Observation number

Example: See 2.13.4

```
> y = c("abc", "def", "ghi")
> x = 1:length(y)
> x
[1] 1 2 3
```

The `length()` function returns the number of elements in a vector. This can be used in conjunction with the `:` operator (Section 2.11.5) to create a vector with the integers from 1 to the sample size. Observation numbers might also be set as case labels as opposed to the row number (see `names()`).

2.4.20 Unique values

Example: See 2.13.4

```
uniquevalues = unique(x)
uniquevalues = unique(data.frame(x1, ..., xk))
```

The `unique()` function returns each of the unique values represented by the vector or dataframe denoted by x (see also `duplicated()`).

2.4.21 Duplicated values

```
isdup = duplicated(x)
```

The `duplicated()` function returns a logical vector consisting of trues if the value remains unique, and false if it has already been observed.

2.4.22 Lagged variable

A lagged variable has the value of that variable in a previous row (typically the immediately previous one) within that dataset. The value of lag for the first observation will be missing (see 2.4.18).

```
lag1 = c(NA, x[1:(length(x)-1)])
```

This expression creates a one-observation lag, with a missing value in the first position, and the first through second to last observation for the remaining entries. We can write a function to create lags of more than one observation.

```
lagk = function(x, k) {
    len = length(x)
    if (!floor(k)==k) {
        cat("k must be an integer")
    } else if (k<1 | k>(len-1)) {
        cat("k must be between 1 and length(x)-1")
    } else {
        return(c(rep(NA, k), x[1:(len-k)]))
    }
}


> lagk(1:10, 5)
 [1] NA NA NA NA NA  1  2  3  4  5
```

2.4.23   SQL

Structured Query Language (SQL) is used to query and modify databases. Access to SQL is available through the RMySQL, RSQLite, or sqldf packages.

2.5   Merging, combining, and subsetting datasets

A common task in data analysis involves the combination, collation, and subsetting of datasets. In this section, we review these techniques for a variety of situations.

2.5.1   Subsetting observations

Example: See 2.13.6

```
smallds = ds[ds$x==1,]
```

This example creates a subject of a dataframe consisting of observations where $X = 1$. In addition, many functions allow specification of a subset=expression option to carry out a procedure on observations that match the expression (see 6.6.8).

2.5.2   Random sample of a dataset

Example: See 7.8.1

It is sometimes useful to sample a subset (here quantified as *nsamp*) of observations with or without replacement from a larger dataset (see also 2.10.11, random number seed).

```
# permutation of a variable
newx = sample(x, replace=FALSE)

# permutation of a dataset
n = dim(ds)[1]  # number of observations
obs = sample(1:n, n, replace=FALSE)
newds = ds[obs,]
```

By default, the `sample()` function takes a sample of all values (determined in this case by determining the number of observations in `ds`), without replacement. This is equivalent to a permutation of the order of values in the vector. The `replace=TRUE` option can be used to override this (e.g., when bootstrapping, see Section 3.1.9). Fewer values can be sampled by specifying the `size` option.

2.5.3   Convert from wide to long (tall) format

See also Section 2.5.4 (reshape from tall to wide)   *Example:* See 5.7.10

Sometimes data are available in a different shape than that required for analysis. One example of this is commonly found in repeated longitudinal measures studies. In this setting it is convenient to store the data in a wide or multivariate format with one line per subject, containing subject invariant factors (e.g., baseline gender), as well as a column for each repeated outcome. An example would be:

```
id female inc80 inc81 inc82
1   0     5000  5500  6000
2   1     2000  2200  3300
3   0     3000  2000  1000
```

where the income for 1980, 1981, and 1982 are included in one row for each id.

In contrast, functions that support repeated measures analyses (5.2.2) typically require a row for each repeated outcome, such as

```
id year female inc
1  80   0      5000
1  81   0      5500
1  82   0      6000
2  80   1      2000
2  81   1      2200
2  82   1      3300
3  80   0      3000
3  81   0      2000
3  82   0      1000
```

In this section and in Section 2.5.4 below, we show how to convert between these two forms of this example data.

```
long = reshape(wide, idvar="id", varying=list(names(wide)[3:5]),
    v.names="inc", timevar="year", times=80:82, direction="long")
```

The list of variables to transpose is provided in the list `varying`, creating `year` as the time variable with values specified by `times` (more flexible dataset transformations are supported by `library(reshape)`).

2.5.4   Convert from long (tall) to wide format

See also Section 2.5.3 (reshape from wide to tall)   *Example:* See 5.7.10

```
wide = reshape(long, v.names="inc", idvar="id", timevar="year",
    direction="wide")
```

This example assumes that the dataset `long` has repeated measures on `inc` for subject `id` determined by the variable `year`. See also `library(reshape)` for more flexible dataset transformations.

2.5.5   Concatenate datasets

```
newds = rbind(ds1, ds2)
```

The result of `rbind()` is a dataframe with as many rows as the sum of rows in `ds1` and `ds2` (see also 2.9.2). Dataframes given as arguments to `rbind()` must have the same column names. The similar `cbind()` function makes a dataframe with as many columns as the sum of the columns in the input objects.

2.5.6   Sort datasets

Example: See 2.13.6

```
sortds = ds[order(x1, x2, ..., xk),]
```

The command `sort()` can be used to sort a vector, while `order()` can be used to determine the order needed to sort a particular vector. The `decreasing` option can be used to change the default sort order (for all variables). The command `sort(x)` is equivalent to `x[order(x)]`. As an alternative, the sort order of a numeric variable can be reversed by specifying `-x1` instead of `x1`.

2.5.7   Merge datasets

Example: See 5.7.12

Merging datasets is commonly required when data on single units are stored in multiple tables or datasets. We consider a simple example where variables id, year, female and inc are available in one dataset, and variables id and maxval in a second. For this simple example, with the first dataset given as:

```
id year female inc
1  80  0      5000
1  81  0      5500
1  82  0      6000
2  80  1      2000
2  81  1      2200
2  82  1      3300
3  80  0      3000
3  81  0      2000
3  82  0      1000
```

and the second given below.

```
id maxval
2  2400
1  1800
4  1900
```

The desired merged dataset would look like:

```
   id year female  inc maxval
1   1   81       0 5500   1800
2   1   80       0 5000   1800
3   1   82       0 6000   1800
4   2   82       1 3300   2400
5   2   80       1 2000   2400
6   2   81       1 2200   2400
7   3   82       0 1000    NA
8   3   80       0 3000    NA
9   3   81       0 2000    NA
10  4   NA      NA   NA   1900
```

```
newds = merge(ds1, ds2, by=id, all=TRUE)
```

The all option specifies that extra rows will be added to the output for any rows that have no matches in the other dataset. Multiple variables can be specified in the by option; if this is left out all variables in both datasets are used: see help(merge).

2.5.8   Drop variables in a dataset

Example: See 2.13.1

It is often desirable to prune extraneous variables from a dataset to simplify analyses.

```
ds[,c("x1", "xk")]
```

The above example created a new dataframe consisting of the variables x1 and xk. An alternative is to specify the variables to be excluded (in this case the second):

```
ds[,names(ds)[-2]]
```

or

```
ds[,-2]
```

More sophisticated ways of listing the variables to be kept are available. For example, the command `ds[,grep("x1|^pat", names(ds))]` would keep x1 and all variables starting with `pat` (see also 2.4.7).

2.6   Date and time variables

The date functions return a `Date` class that represents the number of days since January 1, 1970. The function `as.numeric()` can be used to create a numeric variable with the number of days since 1/1/1970 (see also the `chron` package).

2.6.1   Create date variable

```
dayvar = as.Date("2010-04-29")
todays_date = as.Date(Sys.time())
```

The return value of `as.Date()` is a `Date` class object. If converted to numeric dayvar it represents the number of days between January 1, 1970 and April 29, 2010, while todays_date is the integer number of days since January 1, 1970 (see also 2.1.4).

2.6.2   Extract weekday

```
wkday = weekdays(datevar)
```

The variable `wkday` contains a string with the English name (e.g., `"Tuesday"`) of the weekday of the `datevar` object.

2.6.3 Extract month

```
monthval = months(datevar)
```

The function `months()` returns a string with the English name of the month (e.g., `"April"`) of the `datevar` object.

2.6.4 Extract year

```
yearval = substr(as.POSIXct(datevar), 1, 4)
```

The `as.POSIXct()` function returns a string representing the date, with the first four characters corresponding to the year.

2.6.5 Extract quarter

```
qrtval = quarters(datevar)
```

The function `quarters()` returns a string representing the quarter of the year (e.g., `"Q1"` or `"Q2"`) given by the `datevar` object.

2.6.6 Create time variable

See also 2.7.1 (timing commands)

```
> arbtime = as.POSIXlt("2010-04-29 17:15:45 NZDT")
> arbtime
[1] "2010-04-29 17:15:45"
> Sys.time()
[1] "2010-04-01 10:12:11 EST"
```

Two time objects can be compared with the subtraction operator to monitor elapsed time.

2.7 Interactions with the operating system

2.7.1 Timing commands

```
system.time({expression})
```

The `expression` (e.g., call to any user or system defined function, see 1.5.1) given as argument to the `system.time()` function is evaluated, and the user, system, and total (elapsed) time is returned. Note that if the expression includes assignment using the = operator, the expression must be enclosed in braces as denoted above.

2.7.2 Suspend execution for a time interval

```
Sys.sleep(numseconds)
```

The command `Sys.sleep()` will pause execution for `numseconds`, with minimal system impact.

2.7.3 Execute command in operating system

```
system("ls")     # Mac OS X and Linux
shell("dir")     # Windows
```

The command `ls` lists the files in the current working directory under most operating systems (see 2.7.7 to capture this information). The `shell()` command can be used under Windows.

2.7.4 Command history

```
savehistory()
loadhistory()
history()
```

The command `savehistory()` saves the history of recent commands, which can be reloaded using `loadhistory()` or displayed using `history()`. The `timestamp()` function can be used to add a date and time stamp to the history.

2.7.5 Find working directory

```
getwd()
```

The command `getwd()` returns the current working directory.

2.7.6   Change working directory

```
setwd("dir_location")
```

The command `setwd()` changes the current working directory to the (absolute or relative) pathname given as argument (and silently returns the current directory). Directory changes can also be done interactively under Windows and Mac OS X by selecting the `Change Working Directory` option under the `Misc` menu.

2.7.7   List and access files

```
list.files()
```

The `list.files()` command returns a character vector of filenames in the current directory (by default). Recursive listings are also supported. The function `file.choose()` provides an interactive file browser, and can be given as an argument to functions such as `read.table()` (Section 2.1.2) or `read.csv()` (Section 2.1.5). Related file operation functions include `file.access()`, `file.info()`, and `files()`.

2.8   Mathematical functions

2.8.1   Basic functions

```
minx = min(x)
maxx = max(x)
parallelmin = pmin(x, y)   # maximum of each element from x and y
parallelmax = pmin(x, y)   # minimum of each element from x and y
meanx = mean(x)
stddevx = sd(x)
absolutevaluex = abs(x)
squarerootx = sqrt(x)
etothex = exp(x)
xtothey = x^y
naturallogx = log(x)
logbase10x = log10(x)
logbase2x = log2(x)
logbasearbx = log(x, base=42)
```

Some functions (e.g., `sd()`) operate on a column basis if given a matrix as argument. The `colwise()` function in `library(plyr)` can be used to turn a function that operates on a vector into a function that operates column-wise on a dataframe (see also `apply()`).

2.8.2 Trigonometric functions

```
sin(pi)
cos(0)
tan(pi/4)
acos(x)
asin(x)
atan(x)
atan2(x, y)
```

2.8.3 Special functions

```
betaxy = beta(x, y)
gammax = gamma(x)
factorialn = factorial(n)
nchooser = choose(n, r)

library(gtools)
nchooser = length(combinations(n, r)[,1])
npermr = length(permutations(n, r)[,1])
```

The `combinations()` and `permutations()` functions return a list of possible combinations and permutations.

2.8.4 Integer functions

See also 2.2.8 (rounding and number of digits to display)

```
nextint = ceiling(x)
justint = floor(x)
round2dec = round(x, 2)
roundint = round(x)
keep4sig = signif(x, 4)
movetozero = trunc(x)
```

The second parameter of the `round()` function determines how many decimal places to round (see also `as.integer()`). The value of `movetozero` is the same as `justint` if $x > 0$ or `nextint` if $x < 0$.

2.8.5 Comparisons of floating point variables

Because certain floating point values of variables do not have exact decimal equivalents, there may be some error in how they are represented on a computer. For example, if the true value of a particular variable is 1/7, the approximate decimal is 0.1428571428571428. For some operations (for example, tests of equality), this approximation can be problematic.

```
> all.equal(0.1428571, 1/7)
[1] "Mean relative difference: 3.000000900364093e-07"
> all.equal(0.1428571, 1/7, tolerance=0.0001)
[1] TRUE
```

The tolerance option for the `all.equal()` function determines how many decimal places to use in the comparison of the vectors or scalars (the default tolerance is set to the underlying lower level machine precision).

2.8.6 Complex numbers

Support for complex numbers is available.

```
(0+1i)^2
```

The above expression is equivalent to $i^2 = -1$. Additional support is available through the `complex()` function and related routines.

2.8.7 Derivative

Rudimentary support for finding derivatives is available. These functions are particularly useful for high-dimensional optimization problems (see 2.8.8).

```
D(expression(x^3), "x")
```

Second (or higher order) derivatives can be found by repeatedly applying the D function with respect to X. This function (as well as `deriv()`) are useful in numerical optimization (see the `nlm()`, `optim()` and `optimize()` functions).

2.8.8 Optimization problems

R can be used to solve optimization (maximization) problems. As an extremely simple example, consider maximizing the area of a rectangle with perimeter equal to 20. Each of the sides can be represented by x and 10-x, with area of the rectangle equal to $x * (10 - x)$.

```
f = function(x) { return(x*(10-x)) }
optimize(f, interval=c(0, 10), maximum=TRUE)
```

Other optimization functions available include `nlm()`, `uniroot()`, `optim()`, and `constrOptim()` (see also the CRAN Optimization and Mathematical Programming Task View) and the `lpSolve` package for linear/integer problems.

2.9   Matrix operations

Matrix operations are often needed to implement statistical methods. Matrices can be created using the `matrix()` function (see 1.5.5 or below). In addition to the routines described below, the `Matrix` library is particularly useful for manipulation of large as well as sparse matrices.

Throughout this section, we use capital letters to emphasize that a matrix is described.

2.9.1   Create matrix from vector

In this entry, we demonstrate creating a 2×2 matrix.

$$A = \begin{pmatrix} 1 & 2 \\ 3 & 4 \end{pmatrix}.$$

```
A = matrix(c(1, 2, 3, 4), 2, 2, byrow=TRUE)
```

2.9.2   Create matrix from vectors

In this entry, we demonstrate creating a $n \times k$ matrix from a set of column vectors each of length n (see also 4.5.9, design and information matrix).

```
A = cbind(x1, ..., xk)
```

This also works to combine appropriately dimensioned matrices, and can be done using rows with the function `rbind()` (see also 2.5.5).

2.9.3   Transpose matrix

```
A = matrix(c(1, 2, 3, 4), 2, 2, byrow=TRUE)
transA = t(A)
```

2.9.4   Find dimension of a matrix

```
A = matrix(c(1, 2, 3, 4), 2, 2, byrow=TRUE)
dim(A)
```

The `dim()` function returns the dimension (number of rows and columns) for rectangular objects (e.g., matrices and dataframes), as opposed to the `length()` function, which operates on vectors.

2.9.5   Matrix multiplication

```
A = matrix(c(1, 2, 3, 4), 2, 2, byrow=TRUE)
Asquared = A %*% A
```

2.9.6   Component-wise multiplication

```
A = matrix(c(1, 2, 3, 4), 2, 2, byrow=TRUE)
newmat = A * A
```

Unlike the matrix multiplication in 2.9.5, the result of this operation is scalar multiplication of each element in the matrix A, which yields:

$$newmat = \begin{pmatrix} 1 & 4 \\ 9 & 16 \end{pmatrix}.$$

2.9.7   Invert matrix

```
A = matrix(c(1, 2, 3, 4), 2, 2, byrow=TRUE)
Ainv = solve(A)
```

2.9.8   Create submatrix

```
A = matrix(1:12, 3, 4, byrow=TRUE)
Asub = A[2:3, 3:4]
```

2.9.9   Create a diagonal matrix

```
A = matrix(c(1, 2, 3, 4), 2, 2, byrow=TRUE)
diagMat = diag(c(1, 4))     # argument is a vector
diagMat = diag(diag(A))     # A is a matrix
```

For vector argument, the `diag()` function generates a matrix with the vector values as the diagonals and all off-diagonals 0. For matrix A, the `diag()` function creates a vector of the diagonal elements (see 2.9.10); a diagonal matrix with these diagonal entries, but all off-diagonals set to 0 can be created by running the `diag()` with this vector as argument.

2.9.10   Create vector of diagonal elements

```
A = matrix(c(1, 2, 3, 4), 2, 2, byrow=TRUE)
diagVals = diag(A)
```

2.9.11   Create vector from a matrix

```
A = matrix(c(1, 2, 3, 4), 2, 2, byrow=TRUE)
newvec = c(A)
```

2.9.12   Calculate determinant

```
A = matrix(c(1, 2, 3, 4), 2, 2, byrow=TRUE)
detval = det(A)
```

2.9.13   Find eigenvalues and eigenvectors

```
A <- matrix(c(1, 2, 3, 4), 2, 2, byrow=TRUE)
Aev <- eigen(A)
Aeval <- Aev$values
Aevec <- Aev$vectors
```

The `eigen()` function in R returns a list consisting of the eigenvalues and eigenvectors, respectively, of the matrix given as argument.

2.9.14   Calculate singular value decomposition

Example: See 2.10.7

The singular value decomposition of a matrix A is given by $A = U * \mathrm{diag}(Q) * V^T$ where $U^T U = V^T V = V V^T = I$ and Q contains the singular values of A.

```
A = matrix(c(1, 2, 3, 4), 2, 2, byrow=TRUE)
svdres = svd(A)
U = svdres$u
Q = svdres$d
V = svdres$v
```

The svd() function returns a list with components corresponding to a vector of singular values, a matrix with columns corresponding to the left singular values, and a matrix with columns containing the right singular values.

2.10   Probability distributions and random number generation

R can calculate quantiles and cumulative distribution values as well as generate random numbers for a large number of distributions. Random variables are commonly needed for simulation and analysis.

A seed can be specified for the random number generator. This is important to allow replication of results (e.g., while testing and debugging). Information about random number seeds can be found in Section 2.10.11.

Table 2.1 summarizes support for quantiles, cumulative distribution functions, and random numbers. Prepend d to the command to compute quantiles of a distribution dNAME(xvalue, parm1, ..., parmn), p for the cumulative distribution function, pNAME(xvalue, parm1, ..., parmn), q for the quantile function qNAME(prob, parm1, ..., parmn), and r to generate random variables rNAME(nrand, parm1, ..., parmn) where in the last case a vector of nrand values is the result.

More information on probability distributions can be found in the CRAN Probability Distributions Task View.

2.10.1   Probability density function

Example: See 2.13.7

Here we use the normal distribution as an example; others are shown in Table 2.1 on the next page.

```
> dnorm(1.96, mean=0, sd=1)
[1] 0.05844094
> dnorm(0, mean=0, sd=1)
[1] 0.3989423
```

2.10.2   Cumulative density function

Here we use the normal distribution as an example; others are shown in Table 2.1.

```
> pnorm(1.96, mean=0, sd=1)
[1] 0.9750021
```

Table 2.1: Quantiles, Probabilities, and Pseudorandom Number Generation: Available Distributions

Distribution	NAME
Beta	beta
Beta-binomial	betabin*
binomial	binom
Cauchy	cauchy
chi-square	chisq
exponential	exp
F	f
gamma	gamma
geometric	geom
hypergeometric	hyper
inverse normal	inv.gaussian*
Laplace	laplace*
logistic	logis
lognormal	lnorm
negative binomial	nbinom
normal	norm
Poisson	pois
Student's t	t
Uniform	unif
Weibull	weibull

Note: See Section 2.10 for details regarding syntax to call these routines.
* The `betabin()`, `inv.gaussian()`, and `laplace()` families of distributions are available using `library(VGAM)`.

2.10.3   Quantiles of a probability density function

Here we calculate the upper 97.5% percentile of the normal distribution as an example; others are shown in Table 2.1.

```
> qnorm(.975, mean=0, sd=1)
[1] 1.959964
```

2.10.4   Uniform random variables

```
x = runif(n, min=0, max=1)
```

The arguments specify the number of variables to be created and the range over which they are distributed (by default unit interval).

2.10.5   Multinomial random variables

```
x = sample(1:r, n, replace=TRUE, prob=c(p1, p2, ..., pr))
```

Here $\sum_r p_r = 1$ are the desired probabilities (see also `rmultinom()` in the `stats` package as well as the `cut()` function).

2.10.6   Normal random variables

Example: See 2.13.7

```
x1 = rnorm(n)
x2 = rnorm(n, mean=mu, sd=sigma)
```

The arguments specify the number of variables to be created and (optionally) the mean and standard deviation (default $\mu = 0$ and $\sigma = 1$).

2.10.7   Multivariate normal random variables

For the following, we first create a 3×3 covariance matrix. Then we generate 1000 realizations of a multivariate normal vector with the appropriate correlation or covariance.

```
library(MASS)
mu = rep(0, 3)
Sigma = matrix(c(3, 1, 2,
                 1, 4, 0,
                 2, 0, 5), nrow=3)
xvals = mvrnorm(1000, mu, Sigma)
apply(xvals, 2, mean)
```

or

```
rmultnorm = function(n, mu, vmat, tol=1e-07)
# a function to generate random multivariate Gaussians
{
   p = ncol(vmat)
   if (length(mu)!=p)
      stop("mu vector is the wrong length")
   if (max(abs(vmat - t(vmat))) > tol)
      stop("vmat not symmetric")
   vs = svd(vmat)
   vsqrt = t(vs$v %*% (t(vs$u) * sqrt(vs$d)))
   ans = matrix(rnorm(n * p), nrow=n) %*% vsqrt
   ans = sweep(ans, 2, mu, "+")
   dimnames(ans) = list(NULL, dimnames(vmat)[[2]])
   return(ans)
}
xvals = rmultnorm(1000, mu, Sigma)
apply(xvals, 2, mean)
```

The returned object `xvals`, of dimension 1000×3, is generated from the variance covariance matrix denoted by `Sigma`, which has first row and column (3,1,2). An arbitrary mean vector can be specified using the `c()` function.

Several techniques are illustrated in the definition of the `rmultnorm()` function. The first lines test for the appropriate arguments, and return an error (see 2.11.3) if the conditions are not satisfied. The singular value decomposition (see 2.9.14) is carried out on the variance covariance matrix, and the `sweep()` function is used to transform the univariate normal random variables generated by `rnorm()` to the desired mean and covariance. The `dimnames()` function applies the existing names (if any) for the variables in `vmat`, and the result is returned.

2.10.8   Truncated multivariate normal random variables

```
library(tmvtnorm)
x = rtmvnorm(n, mean, sigma, lower, upper)
```

The arguments specify the number of variables to be created, the mean and standard deviation, and a vector of lower and upper truncation values.

2.10.9   Exponential random variables

```
x = rexp(n, rate=lambda)
```

The arguments specify the number of variables to be created and (optionally) the inverse of the mean (default $\lambda = 1$).

2.10.10   Other random variables

Example: See 2.13.7

The list of probability distributions supported can be found in Table 2.1. In addition to these distributions, the inverse probability integral transform can be used to generate arbitrary random variables with invertible cumulative density function F (exploiting the fact that $F^{-1} \sim U(0,1)$). As an example, consider the generation of random variates from an exponential distribution with rate parameter λ, where $F(X) = 1 - \exp(-\lambda X) = U$. Solving for X yields $X = -\log(1 - U)/\lambda$. If we generate 500 Uniform(0,1) variables, we can use this relationship to generate 500 exponential random variables with the desired rate parameter (see also 7.3.4, sampling from pathological distributions).

```
lambda = 2
expvar = -log(1-runif(500))/lambda
```

2.10.11   Setting the random number seed

The default behavior is a (pseudorandom) seed based on the system clock. To generate a replicable series of variates, first run `set.seed(seedval)` where `seedval` is a single integer for the default "Mersenne-Twister" random number generator. For example:

```
set.seed(42)
set.seed(Sys.time())
```

More information can be found using `help(.Random.seed)`.

2.11   Control flow, programming, and data generation

Here we show some basic aspects of control flow, programming, and data generation (see also 7.2, data generation and 1.6.2, writing functions).

2.11.1   Looping

Example: See 7.1.2

```
x = numeric(i2-i1+1)    # create placeholder
for (i in 1:length(x)) {
    x[i] = rnorm(1) # this is slow and inefficient!
}
```

or (preferably)

```
x = rnorm(i2-i1+1)   # this is far better
```

Most tasks that could be written as a loop are often dramatically faster if they are encoded as a vector operation (as in the second and preferred option above). Examples of situations where loops are particularly useful can be found in Sections 4.1.6 and 7.1.2. More information on control structures for looping and conditional processing can be found in `help(Control)`.

2.11.2   Error recovery

Example: See 2.13.2

```
try(expression, silent=FALSE)
```

The `try()` function runs the given `expression` and traps any errors that may arise (displaying them on the standard error output device). The related function `geterrmessage()` can be used to display any errors.

2.11.3   Assertions

Example: See 2.10.7 and 2.13.2

Assertions can be useful in data consistency checking and defensive coding.

```
stopifnot(expr1, ..., exprk)
```

The `stopifnot()` function runs the given expressions and returns an error message if all are not true (see also `stop()`). As an example, we can consider two equivalent ways to test that the variable `age` is non-negative.

```
stopifnot(age>=0)
```

or

```
if (sum(age<0)>=1) stop("at least one age is negative!")
```

2.11.4   Conditional execution

Example: See 4.7.6 and 6.6.9

```
if (expression1) { expression2 }
```

or

```
if (expression1) { expression2 } else { expression3 }
```

or

```
ifelse(expression, x, y)
```

The `if` statement, with or without `else`, tests a single logical statement; it is not an elementwise (vector) function. If `expression1` evaluates to `TRUE`, then `expression2` is evaluated. The `ifelse()` function operates on vectors and evaluates the expression given as `expression` and returns `x` if it is `TRUE` and `y` otherwise (see also comparisons, 1.5.2). An expression can include multi-command blocks of code (in brackets).

2.11.5   Sequence of values or patterns

Example: See 2.13.7

It is often useful to generate a variable consisting of a sequence of values (e.g., the integers from 1 to 100) or a pattern of values (1 1 1 2 2 2 3 3 3). This might be needed to generate a variable consisting of a set of repeated values for use in a simulation or graphical display.

```
# generate
seq(from=i1, to=i2, length.out=nvals)
seq(from=i1, to=i2, by=1)
seq(i1, i2)
i1:i2

rep(value, times=nvals)
```
or
```
rep(value, each=nvals)
```

The `seq` function creates a vector of length `nvals` if the `length.out` option is specified. If the `by` option is included, the length is approximately `(i2-i1)/byval`. The `i1:i2` operator is equivalent to `seq(from=i1, to=i2, by=1)`. The `rep` function creates a vector of length `nvals` with all values equal to `value`, which can be a scalar, vector, or list. The `each` option repeats each element of `value` `nvals` times. The default is `times`.

As an example, we demonstrate generating data from a linear regression model (4.1.1) with normal errors of the form:

$$E[Y|X_1, X_2] = \beta_0 + \beta_1 X_1 + \beta_2 X_2, \ Var(Y|X) = 3, \ Corr(X_1, X_2) = 0.$$

The following code implements the model described above for $n = 2000$. The `table()` function (see 3.2.2) is used to check whether the intended distribution of covariates was achieved, and the `coef()` and `lm()` functions can fit a model using these simulated data.

```
> n = 2000
> x1 = rep(c(0,1), each=n/2)      # x1 resembles 0 0 0 ... 1 1 1
> x2 = rep(c(0,1), times=n/2)     # x2 resembles 0 1 0 1 ... 0 1
> beta0 = -1; beta1 = 1.5; beta2 = .5;
> mse = 3
> table(x1, x2)
    x2
x1   0    1
  0 500 500
  1 500 500
> y = beta0 + beta1*x1 + beta2*x2 + rnorm(n, 0, mse)
> coef(lm(y ~ x1 + x2))
(Intercept)          x1            x2
    -0.965        1.433         0.537
```

2.11.6 Grid of values

Example: See 7.8.2

It is straightforward to generate a dataframe with all combinations of two or more vectors.

```
> expand.grid(x1=1:3, x2=c("M", "F"))
  x1 x2
1  1  M
2  2  M
3  3  M
4  1  F
5  2  F
6  3  F
```

The `expand.grid()` function takes two or more vectors or factors and returns a data frame. The first factors vary fastest.

2.11.7 Reference a variable using a character vector

Example: See 5.7.10

A variable can be referenced using a character vector rather than its actual name, which can be helpful when programming.

```
mean(x)
mean(get("x"))
```

The `get()` function searches for an object within the workspace, and returns it. If there is a variable x in the workspace, both of these commands will calculate its mean. This can be useful when accessing elements within lists, as they can be referenced symbolically as well.

```
> newlist = list(x1=3, x2="Yes", x3=TRUE)
> newlist[[2]]
[1] "Yes"
> val="x2"
> newlist[[val]]
[1] "Yes"
```

2.11.8 Perform an action repeatedly over a set of variables

Example: See 2.13.5 and 5.7.10

It is often necessary to perform a given function for a series of variables. Here the square of each of a list of variables is calculated as an example.

```
l1 = c("x1", "x2", ..., "xk")
l2 = c("z1", "z2", ..., "zk")
for (i in 1:length(l1)) {
    assign(l2[i], eval(as.name(l1[i]))^2)
}
```

It is not straightforward to refer to objects without evaluating those objects. Assignments objects given symbolically can be made using the `assign()` function. Here a somewhat obscure use of the `eval()` function is used to evaluate an expression after the string value in `l1` is coerced to be a symbol. This allows the values of the character vectors `l1` and `l2` to be evaluated (see `get()`, `help(assign)` and `help(eval)`).

2.12   Further resources

A comprehensive review of introductory statistics in R is accessibly presented by Verzani [77]. Paul Murrell's *Introduction to Data Technologies* text [46] provides a comprehensive introduction to XML, SQL, and other related technologies and can be found at `http://www.stat.auckland.ac.nz/~paul/ItDT`.

2.13   HELP examples

To help illustrate the tools presented in this chapter, we apply many of the entries to the HELP data. The code for these examples can be downloaded from `http://www.math.smith.edu/r/examples`.

2.13.1   Data input

We begin by reading the dataset (2.1.5), keeping only the variables that are needed (2.5.8).

```
> options(digits=3)
> options(width=68) # narrow output
> ds = read.csv("http://www.math.smith.edu/r/data/help.csv")
> newds = ds[,c("cesd","female","i1","i2","id","treat","f1a",
+     "f1b","f1c","f1d","f1e","f1f","f1g","f1h","f1i","f1j","f1k",
+     "f1l","f1m","f1n","f1o","f1p","f1q","f1r","f1s","f1t")]
```

We can then display a summary of the dataset. The default output prints a line for each variable with its name and additional information; the `short` option below limits the output to just the names of the variable.

```
> attach(newds)
> names(newds)

 [1] "cesd"    "female"  "i1"      "i2"      "id"      "treat"   "f1a"
 [8] "f1b"     "f1c"     "f1d"     "f1e"     "f1f"     "f1g"     "f1h"
[15] "f1i"     "f1j"     "f1k"     "f1l"     "f1m"     "f1n"     "f1o"
[22] "f1p"     "f1q"     "f1r"     "f1s"     "f1t"

> # structure of the first 10 variables
> str(newds[,1:10])

'data.frame':            453 obs. of  10 variables:
 $ cesd  : int  49 30 39 15 39 6 52 32 50 46 ...
 $ female: int  0 0 0 1 0 1 1 0 1 0 ...
 $ i1    : int  13 56 0 5 10 4 13 12 71 20 ...
 $ i2    : int  26 62 0 5 13 4 20 24 129 27 ...
 $ id    : int  1 2 3 4 5 6 7 8 9 10 ...
 $ treat : int  1 1 0 0 0 1 0 1 0 1 ...
 $ f1a   : int  3 3 3 0 3 1 3 1 3 2 ...
 $ f1b   : int  2 2 2 0 0 0 1 1 2 3 ...
 $ f1c   : int  3 0 3 1 3 1 3 2 3 3 ...
 $ f1d   : int  0 3 0 3 3 3 1 3 1 0 ...
```

We can also display the means of the first 10 variables.

```
> library(plyr)
> colwise(mean)(newds[,1:10])

  cesd female   i1   i2  id treat  f1a  f1b  f1c  f1d
1 32.8  0.236 17.9 22.6 233 0.497 1.63 1.39 1.92 1.57
```

Displaying the first few rows of data can give a more concrete sense of what is in the dataset.

```
> head(newds, n=5)

  cesd female i1 i2 id treat f1a f1b f1c f1d f1e f1f f1g f1h f1i
1   49      0 13 26  1     1   3   2   3   0   2   3   3   0   2
2   30      0 56 62  2     1   3   2   0   3   3   2   0   0   3
3   39      0  0  0  3     0   3   2   3   0   2   2   1   3   2
4   15      1  5  5  4     0   0   0   1   3   2   2   1   3   0
5   39      0 10 13  5     0   3   0   3   3   3   3   1   3   3
  f1j f1k f1l f1m f1n f1o f1p f1q f1r f1s f1t
1   3   3   0   1   2   2   2   2   3   3   2
2   0   3   0   0   3   0   0   0   2   0   0
3   3   1   0   1   3   2   0   0   3   2   0
4   0   1   2   2   2   0  NA   2   0   0   1
5   2   3   2   2   3   0   3   3   3   3   3
```

2.13.2 Consistency checking

Finally, we can do a consistency check using a series of assertions (2.11.3).

```
> stopifnot(cesd>=0 & cesd<=60, i1>=0, female==0 | female==1)
```

The function returns nothing, since the argument is TRUE. We can also assert something that is not true (all subjects drank), and trap then display the error.

```
> try(stopifnot(i1>=1))
> geterrmessage()

[1] "Error : i1 >= 1 is not all TRUE\n"
```

2.13.3 Data output

Saving the dataset in native format (2.2.1) will ease future access. We also add a comment (2.3.4) to help later users understand what is in the dataset.

```
> comment(newds) = "HELP baseline dataset"
> comment(newds)

[1] "HELP baseline dataset"

> save(ds, file="savedfile")
```

Saving it in a text format (e.g., comma separated, 2.2.3) to be read into Excel will facilitate transfer.

```
> write.csv(newds, "file.csv")
```

Other file conversions are supported (2.1.6). As an example, to get data from R into SAS, the following code generates an ASCII dataset as well as a SAS command file to read it in to SAS.

```
> library(foreign)
> write.foreign(newds, "file.dat", "file.sas", package="SAS")
```

2.13.4 Data display

We begin by consideration of the CESD (Center for Epidemiologic Studies–Depression) measure of depressive symptoms for this sample at baseline.

The indexing mechanisms in R (see 1.5.2) are helpful in extracting subsets of a vector.

```
> cesd[1:10]
```

```
[1] 49 30 39 15 39  6 52 32 50 46
```

It may be useful to know how many high values there are, and to which observations they belong:

```
> cesd[cesd>55]
```

```
[1] 57 58 57 60 58 56 58 56 57 56
```

```
> # which rows have values this high?
> which(cesd>55)
```

```
[1]   64 116 171 194 231 266 295 305 387 415
```

Similarly, it may be useful to examine the observations with the lowest values.

```
> sort(cesd)[1:4]
```

```
[1] 1 3 3 4
```

2.13.5   Derived variables and data manipulation

Suppose the dataset arrived with only the individual CESD questions, and not the sum (total score). We would need to create the CESD score. Four questions are asked "backwards," meaning that high values of the response are counted for fewer points.[1] We will approach the backwards questions by reading the CESD items into a new object. To demonstrate other tools, we will also see if there are any missing data (2.4.18), and reconstruct how the original creators of the dataset handled missingness.

```
> table(is.na(f1g))
```

```
FALSE  TRUE
  452     1
```

```
> # reverse code f1d, f1h, f1l and f1p
> cesditems = cbind(f1a, f1b, f1c, (3 - f1d), f1e, f1f, f1g,
+     (3 - f1h), f1i, f1j, f1k, (3 - f1l), f1m, f1n, f1o,
+     (3 - f1p), f1q, f1r, f1s, f1t)
```

---

[1]According to the coding instructions at http://patienteducation.stanford.edu/research/cesd.pdf.

```
> nmisscesd = apply(is.na(cesditems), 1, sum)
> ncesditems = cesditems
> ncesditems[is.na(cesditems)] = 0
> newcesd = apply(ncesditems, 1, sum)
> imputemeancesd = 20/(20-nmisscesd)*newcesd
```

It is prudent to review the results when deriving variables. We will check our recreated CESD score against the one which came with the dataset. To ensure that missing data has been correctly coded, we print the subjects with any missing questions.

```
> cbind(newcesd, cesd, nmisscesd, imputemeancesd)[nmisscesd>0,]

     newcesd cesd nmisscesd imputemeancesd
[1,]      15   15         1           15.8
[2,]      19   19         1           20.0
[3,]      44   44         1           46.3
[4,]      17   17         1           17.9
[5,]      29   29         1           30.5
[6,]      44   44         1           46.3
[7,]      39   39         1           41.1
```

The output shows that the original variable was calculated incorporating unanswered questions counted as if they had been answered with a zero. This conforms to the instructions provided with the CESD, but might be questioned on theoretical grounds.

It is often necessary to create a new variable using logic (2.4.15). In the HELP study, many subjects reported extreme amounts of drinking (as the baseline measure was taken while they were in detox). Here, an ordinal measure of alcohol consumption (abstinent, moderate, high-risk) is created using information about average consumption per day in past 30 days prior to detox (i1, measured in standard drink units) and maximum number of drinks per day in past 30 days prior to detox (i2). The number of drinks required for each category differ for men and women according to National Institute of Alcohol Abuse and Alcoholism (NIAAA) guidelines for physicians [48].

```
> # create empty repository for new variable
> drinkstat = character(length(i1))
> # create abstinent group
> drinkstat[i1==0] = "abstinent"
> # create moderate group
> drinkstat[(i1>0 & i1<=1 & i2<=3 & female==1) |
+   (i1>0 & i1<=2 & i2<=4 & female==0)] = "moderate"
> # create highrisk group
> drinkstat[((i1>1 | i2>3) & female==1) |
+   ((i1>2 | i2>4) & female==0)] = "highrisk"
> # do we need to account for missing values?
> is.na(drinkstat) = is.na(i1) | is.na(i2) | is.na(female)
> table(is.na(drinkstat))

FALSE
  453
```

It is always prudent to check the results of derived variables. As a demonstration, we display the observations in the 361st through 370th rows of the data.

```
> tmpds = data.frame(i1, i2, female, drinkstat)
> tmpds[361:370,]

    i1 i2 female drinkstat
361 37 37      0  highrisk
362 25 25      0  highrisk
363 38 38      0  highrisk
364 12 29      0  highrisk
365  6 24      0  highrisk
366  6  6      0  highrisk
367  0  0      0 abstinent
368  0  0      1 abstinent
369  8  8      0  highrisk
370 32 32      0  highrisk
```

We can focus checks on a subset of observations. Here we show the drinking data for moderate female drinkers.

```
> tmpds[tmpds$drinkstat=="moderate" & tmpds$female==1,]

    i1 i2 female drinkstat
116  1  1      1  moderate
137  1  3      1  moderate
225  1  2      1  moderate
230  1  1      1  moderate
264  1  1      1  moderate
266  1  1      1  moderate
394  1  1      1  moderate
```

Now we create a categorical variable from the CESD variable, with values greater than 0 and less than 16 in one group, those between 16 and less than 22 in a second, and those 22 and above in a third (see 2.4.13).

```
> table(cut(cesd, c(0, 16, 22, 60), right=FALSE))

 [0,16) [16,22) [22,60)
     46      42     364
```

Basic data description is an early step in analysis. Here we show some summary statistics related to drinking and gender.

```
> sum(is.na(drinkstat))

[1] 0

> table(drinkstat, exclude="NULL")

drinkstat
abstinent  highrisk  moderate
       68       357        28

> table(drinkstat, female, exclude="NULL")

           female
drinkstat    0   1
  abstinent  42  26
  highrisk  283  74
  moderate   21   7
```

To display gender in a more direct fashion, we create a new variable.

```
> gender = factor(female, c(0,1), c("male","Female"))
> table(female)

female
  0   1
346 107

> table(gender)

gender
  male Female
   346    107
```

2.13.6   Sorting and subsetting datasets

It is often useful to sort datasets (2.5.6) by the order of a particular variable
(or variables). Here we sort by CESD and drinking.

```
> detach(newds)
> newds = ds[order(ds$cesd, ds$i1),]
> newds[1:5,c("cesd", "i1", "id")]

    cesd i1  id
199    1  3 233
394    3  1 139
349    3 13 418
417    4  4 251
85     4  9  95
```

It is sometimes necessary to create data that is a subset (2.5.1) of other data.
For example, here we make a dataset which only includes female subjects. First,
we create the subset and calculate a summary value in the resulting dataset.

```
> females = ds[ds$female==1,]
> attach(females)
> mean(cesd)

[1] 36.9
```

To test the subsetting, we then display the mean for both genders, as described
in Section 3.1.2.

```
> with(ds, mean(cesd[female==1]))

[1] 36.9

> tapply(ds$cesd, ds$female, mean)

   0    1
31.6 36.9

> aggregate(ds$cesd, list(ds$female), mean)

  Group.1    x
1       0 31.6
2       1 36.9
```

2.13.7 Probability distributions

Data can easily be generated. As an example, we can find values of the normal (2.10.6) and t densities, and display them in Figure 2.1.

```
> x = seq(from=-4, to=4.2, length=100)
> normval = dnorm(x, 0, 1)
> dfval = 1
> tval = dt(x, df=dfval)

> plot(x, normval, type="n", ylab="f(x)", las=1)
> lines(x, normval, lty=1, lwd=2)
> lines(x, tval, lty=2, lwd=2)
> legend(1.1, .395, lty=1:2, lwd=2,
+    legend=c(expression(N(mu == 0,sigma == 1)),
+    paste("t with ", dfval," df", sep="")))
> grid(nx=NULL, ny=NULL, col="darkgray")
```

Mathematical symbols (6.2.13) representing the parameters of the normal distribution are included as part of the legend (6.2.15) to help differentiate the distributions. A grid (6.2.7) is also added.

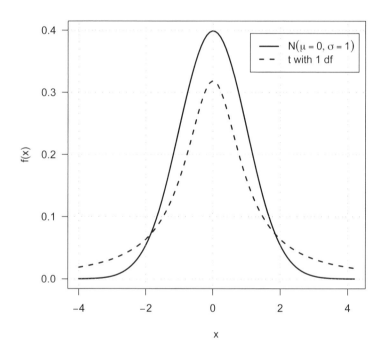

Figure 2.1: Comparison of standard normal and t distribution with 1 df.

Chapter 3

Common statistical procedures

This chapter describes how to generate univariate summary statistics for continuous variables (such as means, variances, and quantiles), display and analyze frequency tables and cross-tabulations for categorical variables, and carry out a variety of one and two sample procedures.

3.1 Summary statistics

3.1.1 Means and other summary statistics

Example: See 3.6.1

```
xmean = mean(x)
```

The `mean()` function accepts a numeric vector or a numeric dataframe as arguments (date objects are also supported). Similar functions include `median()` (see 3.1.5 for more quantiles), `var()`, `sd()`, `min()`, `max()`, `sum()`, `prod()`, and `range()` (note that the latter returns a vector containing the minimum and maximum value). The `which.max()` and `which.min()` functions can be used to identify the observation containing the maximum and minimum values, respectively (see also `which()`, Section 1.5.2).

3.1.2 Means by group

See also 4.1.6 (fitting regression separately by group)   *Example:* See 2.13.6

```
tapply(y, x, mean)
```

or

```
ave(y, as.factor(x), FUN=mean)
```

or

```
aggregate(y, list(x1, x2), mean)
```

The `tapply()` function applies the specified function given as the third argument (in this case `mean()`) to the vector y stratified by every unique set of values of the list of factors specified x (see also 1.6.3, the apply family of functions). It returns a vector with length equal to the number of unique set of values of x. Similar functionality is available using the `ave()` function (see `example(ave)`), which returns a vector of the same length as x with each element equal to the mean of the subset of observations with the factor level specified by y. The `aggregate()` function can be used in a similar manner, with a list of variables given as argument (see also 2.13.6).

3.1.3 Trimmed mean

```
mean(x, trim=frac)
```

The value `frac` can take on range 0 to 0.5, and specifies the fraction of observations to be trimmed from each end of x before the mean is computed (`frac=0.5` yields the median).

3.1.4 Five-number summary

Example: See 3.6.1

The five-number summary (minimum, 25th percentile, median, 75th percentile, maximum) is a useful summary of the distribution of observed values.

```
quantile(x)
fivenum(x)
summary(ds)
```

The `summary()` function calculates the five-number summary (plus the mean) for each of the columns of the vector or dataset given as arguments. The default output of the `quantile()` function is the min, 25th percentile, median, 75th percentile and the maximum. The `fivenum()` function reports the lower and upper hinges instead of the 25th and 75th percentiles, respectively.

3.1.5 Quantiles

Example: See 3.6.1

```
quantile(x, c(.025, .975))
quantile(x, seq(from=.95, to=.975, by=.0125))
```

Details regarding the calculation of quantiles in `quantile()` can be found using `help(quantile)`.

3.1.6 Centering, normalizing, and scaling

```
zscoredx = scale(x)
```

or

```
zscoredx = (x-mean(x))/sd(x)
```

The default behavior of `scale()` is to create a Z-score transformation. The `scale()` function can operate on matrices and dataframes, and allows the specification of a vector of the scaling parameters for both center and scale (see also `sweep()`, a more general function).

3.1.7 Mean and 95% confidence interval

Example: See 3.6.4

```
tcrit = qt(.975, length(x)-1)
ci95 = mean(x) + c(-1,1)*tcrit*sd(x)/sqrt(length(x))
```

or

```
t.test(x)$conf.int
```

While the appropriate 95% confidence interval can be generated in terms of the mean and standard deviation, it is more straightforward to use the t-test function to calculate the relevant quantities.

3.1.8 Maximum likelihood estimation of distributional parameters

Example: See 3.6.1

```
library(MASS)
fitdistr(x, densityfunction}
```

Options for `densityfunction` include `beta`, `cauchy`, `chi-squared`, `exponential`, `f`, `gamma`, `geometric`, `log-normal`, `lognormal`, `logistic`, `negative binomial`, `normal`, `Poisson`, `t` or `weibull`.

3.1.9 Bootstrapping a sample statistic

Bootstrapping is a powerful and elegant approach to estimation of sample statistics that can be implemented even in many situations where asymptotic results are difficult to find or otherwise unsatisfactory [11, 24]. Bootstrapping proceeds using three steps: First, resample the dataset (with replacement) a specified number of times (typically on the order of 10,000), calculate the desired statistic from each resampled dataset, then use the distribution of the resampled statistics to estimate the standard error of the statistic (normal approximation method), or construct a confidence interval using quantiles of that distribution (percentile method).

As an example, we consider estimating the standard error and 95% confidence interval for the coefficient of variation (COV), defined as σ/μ, for a random variable X. Note that for both packages, the user must provide code (as a function) to calculate the statistic of interest.

```
library(boot)
covfun = function(x, i) {sd(x[i])/mean(x[i])}
res = boot(x, covfun, R=10000)
print(res)
plot(res)
quantile(res$t, c(.025, .975))  # percentile method
```

The first argument to the `boot()` function specifies the data to be bootstrapped (in this case a vector, though a dataframe can be set up if more than one variable is needed for the calculation of the sample statistic) as well as a function to calculate the statistic for each resampling iteration. Here the function `covfun()` takes two arguments: The first is the original data (as a vector) and the second a set of indices into that vector (that represent a given bootstrap sample).

The `boot()` function returns an object of class `boot`, with an associated `plot()` function that provides a histogram and QQ-plot (see `help(plot.boot)`). The return value object (`res`, above) contains the vector of resampled statistics (`res$t`), which can be used to estimate the quantiles or standard error. The `boot.ci()` function can be used to generate bias-corrected and accelerated intervals.

3.1.10   Proportion and 95% confidence interval

Example: See 7.1.2

```
binom.test(sum(x), length(x))
prop.test(sum(x), length(x))
```

The `binom.test()` function calculates an exact Clopper–Pearson confidence interval based on the F distribution [4] using the first argument as the number of successes and the second argument the number of trials, while `prop.test()` calculates an approximate confidence interval by inverting the score test. Both allow specification of `p` for the null hypothesis. The `conf.level` option can be used to change the default confidence level.

3.1.11 Tests of normality

```
library(nortest)
ad.test(x)       # Anderson-Darling test
cvm.test(x)      # Cramer-von Mises test
lillie.test(x)   # Lilliefors (KS) test
pearson.test(x)  # Pearson chi-square
sf.test(x)       # Shapiro-Francia test
```

3.2 Contingency tables

3.2.1 Display counts for a single variable

Example: See 3.6.3

Frequency tables display counts of values for a single variable (see also 3.2.2, cross-classification tables).

```
count = table(x)
percent = count/sum(count)*100
rbind(count, percent)
```

Additional marginal displays (in this case the percentages) can be added and displayed along with the counts.

3.2.2 Display cross-classification table

Example: See 3.6.3

Contingency tables display group membership across categorical (grouping) variables. They are also known as cross-classification tables, cross-tabulations, and two-way tables.

```
mytab = table(y, x)
addmargins(mytab)
prop.table(mytab, 1)
```

or

```
xtabs(~ y + x)
```

or

```
library(prettyR)
xtab(y ~ x, data=ds)
```

The `addmargins()` function adds (by default) the row and column totals to a table, while `prop.table()` can be used to calculate row totals (with option 1) and column totals (with option 2). The `colSums()`, `colMeans()` functions (and their equivalents for rows) can be used to efficiently calculate sums and means for numeric arrays. Missing values can be displayed using `table()` by specifying `exclude=NULL`.

The `xtabs()` function can be used to create a contingency table from cross-classifying factors. Much of the process of displaying tables is automated in the `prettyR` library `xtab()` function (which requires specification of a dataframe to operate on).

3.2.3   Pearson chi-square statistic

Example: See 3.6.3

```
chisq.test(x, y)
```

The `chisq.test()` command can accept either two factor vectors or a matrix with counts. By default a continuity correction is used (this can be turned off using the option `correct=FALSE`).

3.2.4   Cochran–Mantel–Haenszel test

The Cochran–Mantel–Haenszel test provides an assessment of the relationship between X_2 and X_3, stratified by (or controlling for) X_1. The analysis provides a way to adjust for the possible confounding effects of X_1 without having to estimate parameters for them.

```
mantelhaen.test(x2, x3, x1)
```

3.2.5   Fisher's exact test

Example: See 3.6.3

```
fisher.test(y, x)
```

or

```
fisher.test(ymat)
```

The `fisher.test()` command can accept either two class vectors or a matrix with counts (here denoted by `ymat`). For tables with many rows and/or columns, p-values can be computed using Monte Carlo simulation using the `simulate.p.value` option.

3.2.6   McNemar's test

McNemar's test tests the null hypothesis that the proportions are equal across matched pairs, for example, when two raters assess a population.

```
mcnemar.test(y, x)
```

The `mcnemar.test()` command can accept either two class vectors or a matrix with counts.

3.3 Bivariate statistics

3.3.1 Epidemiologic statistics

Example: See 3.6.3

It is straightforward to calculate summary measures such as the odds ratio, relative risk and attributable risk (see also 5.1, generalized linear models).

```
sum(x==0&y==0)*sum(x==1&y==1)/(sum(x==0&y==1)*sum(x==1&y==0))
```

or

```
tab1 = table(x, y)
tab1[1,1]*tab1[2,2]/(tab1[1,2]*tab1[2,1])
```

or

```
glm1 = glm(y ~ x, family=binomial)
exp(glm1$coef[2])
```

or

```
library(epitools)
oddsratio.fisher(x, y)
oddsratio.wald(x, y)
riskratio(x, y)
riskratio.wald(x, y)
```

The `epitab()` function in `library(epitools)` provides a general interface to many epidemiologic statistics, while `expand.table()` can be used to create individual level data from a table of counts.

3.3.2 Test characteristics

The sensitivity of a test is defined as the probability that someone with the disease (D=1) tests positive (T=1), while the specificity is the probability that someone without the disease (D=0) tests negative (T=0). For a dichotomous screening measure, the sensitivity and specificity can be defined as $P(D = 1, T = 1)/P(D = 1)$ and $P(D = 0, T = 0)/P(D = 0)$, respectively (see also 6.1.17, receiver operating characteristic curves).

```
sens = sum(D==1&T==1)/sum(D==1)
spec = sum(D==0&T==0)/sum(D==0)
```

Sensitivity and specificity for an outcome D can be calculated for each value of a continuous measure T using the following code.

```
library(ROCR)
pred = prediction(T, D)
diagobj = performance(pred, "sens", "spec")
spec = slot(diagobj, "y.values")[[1]]
sens = slot(diagobj, "x.values")[[1]]
cut = slot(diagobj, "alpha.values")[[1]]
diagmat = cbind(cut, sens, spec)
head(diagmat, 10)
```

The `ROCR` package facilitates the calculation of test characteristics, including sensitivity and specificity. The `prediction()` function takes as arguments the continuous measure and outcome. The returned object can be used to calculate quantities of interest (see `help(performance)` for a comprehensive list). The `slot()` function is used to return the desired sensitivity and specificity values for each cut score, where `[[1]]` denotes the first element of the returned list (see Section 1.5.4, `help(list)`, and `help(Extract)`).

3.3.3   Correlation

Example: See 3.6.2 and 6.6.9

```
pearsoncorr = cor(x, y)
spearmancorr = cor(x, y, method="spearman")
kendalltau = cor(x, y, method="kendall")
```

or

```
cormat = cor(cbind(x1, ..., xk))
```

Tests and confidence intervals for correlations can be generated using the function `cor.test()`. Specifying `method="spearman"` or `method="kendall"` as an option to `cor()` or `cor.test()` generates the Spearman or Kendall correlation coefficients, respectively. A matrix of variables (created with `cbind()`, see 2.9.1) can be used to generate the correlation between a set of variables. Subsets of the returned correlation matrix can be selected, as demonstrated in Section 3.6.2. This can save space by avoiding replicating correlations above and below the diagonal of the correlation matrix. The `use` option for `cor()` specifies how missing values are handled (either `"all.obs"`, `"complete.obs"`, or `"pairwise.complete.obs"`).

3.3.4   Kappa (agreement)

```
library(irr)
kappa2(data.frame(x, y))
```

The `kappa2()` function takes a dataframe (see 1.5.6) as argument. Weights can be specified as an option.

3.4 Two sample tests for continuous variables

3.4.1 Student's t-test

Example: See 3.6.4

```
t.test(y1, y2)
```

or

```
t.test(y ~ x)
```

The first example for the `t.test()` command displays how it can take two vectors (y1 and y2) as arguments to compare, or in the latter example a single vector corresponding to the outcome (y), with another vector indicating group membership (x) using a formula interface (see Sections 1.5.7 and 4.1.1). By default, the two-sample t-test uses an unequal variance assumption. The option `var.equal=TRUE` can be added to specify an equal variance assumption. The command `var.test()` can be used to formally test equality of variances.

3.4.2 Nonparametric tests

Example: See 3.6.4

```
wilcox.test(y1, y2)
ks.test(y1, y2)

library(coin)
median_test(y ~ x)
```

By default, the `wilcox.test()` function uses a continuity correction in the normal approximation for the p-value. The `ks.test()` function does not calculate an exact p-value when there are ties. The median test shown will generate an exact p-value with the `distribution="exact"` option.

3.4.3 Permutation test

Example: See 3.6.4

```
library(coin)
oneway_test(y ~ as.factor(x), distribution=approximate(B=bnum))
```

The `oneway_test` function in the `coin` library implements a variety of permutation-based tests (see also the `exactRankTests` package). An empirical p-value is generated if `distribution=approximate` is specified. This is asymptotically equivalent to the exact p-value, based on `bnum` Monte Carlo replicates.

3.4.4   Logrank test

Example: See 3.6.5

See also 6.1.18 (Kaplan–Meier plot) and 5.3.1 (Cox proportional hazards model)

```
library(survival)
survdiff(Surv(timevar, cens) ~ x)
```

If `cens` is equal to 0, then `Surv()` treats `timevar` as the time of censoring, otherwise the time of the event. Other tests within the G-rho family of Fleming and Harrington [18] are supported by specifying the `rho` option.

3.5   Further resources

Verzani [77] and Everitt and Hothorn [13] present comprehensive introductions for the use of R to fit a common statistical model. Efron and Tibshirani [11] provide a comprehensive overview of bootstrapping. A readable introduction to permutation-based inference can be found in [22]. Collett [5] presents an accessible introduction to survival analysis.

3.6   HELP examples

To help illustrate the tools presented in this chapter, we apply many of the entries to the HELP data. The code for these examples can be downloaded from `http://www.math.smith.edu/r/examples`.

3.6.1   Summary statistics and exploratory data analysis

We begin by reading the dataset.

```
> options(digits=3)
> options(width=68)   # narrows output to stay in the gray box
> ds = read.csv("http://www.math.smith.edu/r/data/help.csv")
> attach(ds)
```

A first step would be to examine some univariate statistics (3.1.1) for the baseline CESD (Center for Epidemiologic Studies–Depression measure of depressive symptoms) score.

We can use functions which produce a set of statistics, such as `fivenum()`, or request them singly.

```
> fivenum(cesd)

[1]  1 25 34 41 60

> mean(cesd); median(cesd)

[1] 32.8

[1] 34

> range(cesd)

[1]  1 60

> sd(cesd)

[1] 12.5

> var(cesd)

[1] 157
```

We can also generate desired statistics. Here, we find the deciles (3.1.5).

```
> quantile(cesd, seq(from=0, to=1, length=11))

  0%  10%  20%  30%  40%  50%  60%  70%  80%  90% 100%
 1.0 15.2 22.0 27.0 30.0 34.0 37.0 40.0 44.0 49.0 60.0
```

Graphics can allow us to easily review the whole distribution of the data. Here we generate a histogram (6.1.9) of CESD, overlaid with its empirical PDF (6.1.21) and the closest-fitting normal distribution (see Figure 3.1).

```
> library(MASS)
> hist(cesd, main="distribution of CESD scores", freq=FALSE)
> lines(density(cesd), lty=2, lwd=2)
> xvals = seq(from=min(cesd), to=max(cesd), length=100)
> param = fitdistr(cesd, "normal")
> lines(xvals, dnorm(xvals, param$estimate[1],
+    param$estimate[2]), lwd=2)
```

3.6.2 Bivariate relationships

We can calculate the correlation (3.3.3) between CESD and MCS and PCS (mental and physical component scores). First, we show the default correlation matrix.

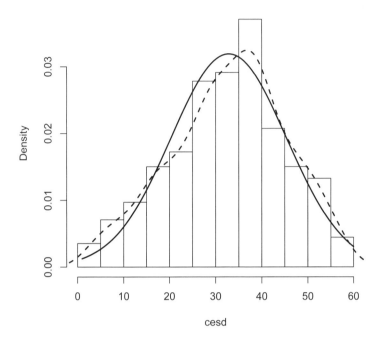

Figure 3.1: Density plot of depressive symptom scores (CESD) plus superimposed histogram and normal distribution.

```
> cormat = cor(cbind(cesd, mcs, pcs))
> cormat

        cesd    mcs    pcs
cesd   1.000 -0.682 -0.293
mcs   -0.682  1.000  0.110
pcs   -0.293  0.110  1.000
```

To save space, we can just print a subset of the correlations.

```
> cormat[c(2, 3), 1]

    mcs    pcs
-0.682 -0.293
```

Figure 3.2 displays a scatterplot (6.1.1) of CESD and MCS, for the female subjects. The plotting character (6.2.1) is the primary substance (Alcohol, Cocaine, or Heroin). We add a rug plot (6.2.9) to help demonstrate the marginal distributions.

```
> plot(cesd[female==1], mcs[female==1], xlab="CESD", ylab="MCS",
+     type="n", bty="n")
> text(cesd[female==1&substance=="alcohol"],
+     mcs[female==1&substance=="alcohol"],"A")
> text(cesd[female==1&substance=="cocaine"],
+     mcs[female==1&substance=="cocaine"],"C")
> text(cesd[female==1&substance=="heroin"],
+     mcs[female==1&substance=="heroin"],"H")
> rug(jitter(mcs[female==1]), side=2)
> rug(jitter(cesd[female==1]), side=3)
```

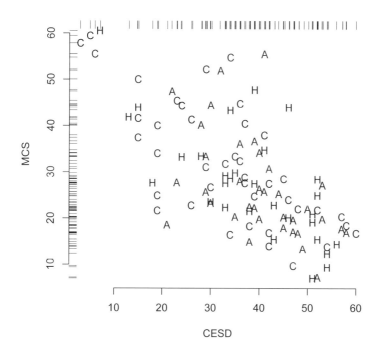

Figure 3.2: Scatterplot of CESD and MCS for women, with primary substance shown as the plot symbol.

3.6.3   Contingency tables

Here we display the cross-classification (contingency) table (3.2.2) of homeless at baseline by gender, calculate the observed odds ratio (OR, Section 3.3.1), and assess association using the Pearson χ^2 test (3.2.3) and Fisher's exact test (3.2.5).

```
> count = table(substance)
> percent = count/sum(count)*100
> rbind(count, percent)

          alcohol  cocaine  heroin
count       177.0    152.0   124.0
percent      39.1     33.6    27.4
```

For cross-classification tables, similar syntax is used.

```
> table(homeless, female)

          female
homeless   0    1
       0 177   67
       1 169   40
```

The prettyR library provides a way to display tables with additional statistics.

```
> library(prettyR)
> xtres = xtab(homeless ~ female, data=ds)

Crosstabulation of homeless by female
           female
homeless        0           1
0             177          67        244
            72.54       27.46      53.86
            51.16       62.62

1             169          40        209
            80.86       19.14      46.14
            48.84       37.38

              346         107        453
            76.38       23.62

odds ratio = 0.63
relative risk (homeless-1) = 0.7
```

We can easily calculate the odds ratio directly.

```
> or = (sum(homeless==0 & female==0)*
+        sum(homeless==1 & female==1))/
+       (sum(homeless==0 & female==1)*
+        sum(homeless==1 & female==0))
> or

[1] 0.625
```

```
> library(epitools)
> oddsobject = oddsratio.wald(homeless, female)
> oddsobject$measure

          odds ratio with 95% C.I.
Predictor estimate lower upper
       0    1.000   NA    NA
       1    0.625 0.401 0.975

> oddsobject$p.value

           two-sided
Predictor midp.exact fisher.exact chi.square
       0        NA           NA          NA
       1     0.0381       0.0456      0.0377
```

The χ^2 and Fisher's exact tests assess independence between gender and homelessness.

```
> chisqval = chisq.test(homeless, female, correct=FALSE)
> chisqval

        Pearson's Chi-squared test

data:  homeless and female
X-squared = 4.32, df = 1, p-value = 0.03767
```

```
> fisher.test(homeless, female)

        Fisher's Exact Test for Count Data

data:  homeless and female
p-value = 0.04560
alternative hypothesis: true odds ratio is not equal to 1
95 percent confidence interval:
 0.389 0.997
sample estimates:
odds ratio
     0.626
```

The `fisher.test()` command returns the conditional MLE for the odds ratio, which is attenuated towards the null value.

3.6.4 Two sample tests of continuous variables

We can assess gender differences in baseline age using a t-test (3.4.1) and non-parametric procedures.

```
> ttres = t.test(age ~ female, data=ds)
> print(ttres)

        Welch Two Sample t-test

data:  age by female
t = -0.93, df = 180, p-value = 0.3537
alternative hypothesis:
true difference in means is not equal to 0

95 percent confidence interval:
 -2.45   0.88
sample estimates:
mean in group 0 mean in group 1
          35.5            36.3
```

The `names()` function can be used to identify the objects returned by the `t.test()` function.

```
> names(ttres)

[1] "statistic"   "parameter"   "p.value"     "conf.int"
[5] "estimate"    "null.value"  "alternative" "method"
[9] "data.name"

> ttres$conf.int

[1] -2.45  0.88
attr(,"conf.level")
[1] 0.95
```

A permutation test (3.4.3) can be run to generate a Monte Carlo p-value.

```
> library(coin)
> oneway_test(age ~ as.factor(female),
+    distribution=approximate(B=9999), data=ds)

        Approximative 2-Sample Permutation Test

data:  age by as.factor(female) (0, 1)
Z = -0.92, p-value = 0.3623
alternative hypothesis: true mu is not equal to 0
```

Similarly, a Wilcoxon nonparametric test (3.4.2) can be requested,

```
> wilcox.test(age ~ as.factor(female), correct=FALSE)

        Wilcoxon rank sum test

data:  age by as.factor(female)
W = 17512, p-value = 0.3979
alternative hypothesis: true location shift is not equal to 0
```

as well as a Kolmogorov–Smirnov test.

```
> ksres = ks.test(age[female==1], age[female==0], data=ds)
> print(ksres)

        Two-sample Kolmogorov-Smirnov test

data:  age[female == 1] and age[female == 0]
D = 0.063, p-value = 0.902
alternative hypothesis: two-sided
```

We can also plot estimated density functions (6.1.21) for age for both groups, and shade some areas (6.2.14) to emphasize how they overlap (Figure 3.3). We create a function (see 1.6) to automate this task.

```
> plotdens = function(x,y, mytitle, mylab) {
+    densx = density(x)
+    densy = density(y)
+    plot(densx, main=mytitle, lwd=3, xlab=mylab, bty="l")
+    lines(densy, lty=2, col=2, lwd=3)
+    xvals = c(densx$x, rev(densy$x))
+    yvals = c(densx$y, rev(densy$y))
+    polygon(xvals, yvals, col="gray")
+ }
```

The polygon() function is used to fill in the area between the two curves.

```
> mytitle = paste("Test of ages: D=", round(ksres$statistic,3),
+     " p=", round(ksres$p.value, 2), sep="")
> plotdens(age[female==1], age[female==0], mytitle=mytitle,
+     mylab="age (in years)")
> legend(50, .05, legend=c("Women", "Men"), col=1:2, lty=1:2,
+     lwd=2)
```

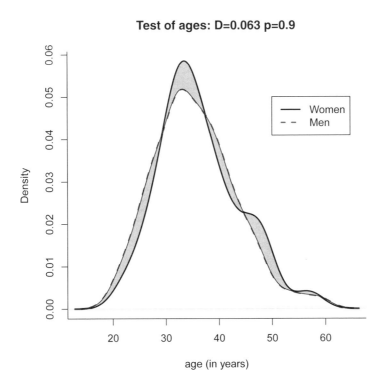

Figure 3.3: Density plot of age by gender.

3.6.5   Survival analysis: Logrank test

The logrank test (3.4.4) can be used to compare estimated survival curves between groups in the presence of censoring. Here we compare randomization groups with respect to `dayslink`, where a value of 0 for `linkstatus` indicates that the observation was censored, not observed, at the time recorded in `dayslink`.

```
> library(survival)
> survobj = survdiff(Surv(dayslink, linkstatus) ~ treat,
+     data=ds)
> print(survobj)

Call:
survdiff(formula = Surv(dayslink, linkstatus) ~ treat, data = ds)

n=431, 22 observations deleted due to missingness.

          N Observed Expected (O-E)^2/E (O-E)^2/V
treat=0 209       35     92.8      36.0      84.8
treat=1 222      128     70.2      47.6      84.8

 Chisq= 84.8  on 1 degrees of freedom, p= 0

> names(survobj)

[1] "n"         "obs"       "exp"       "var"       "chisq"
[6] "na.action" "call"
```

Chapter 4

Linear regression and ANOVA

Regression and analysis of variance (ANOVA) form the basis of many investigations. In this chapter we describe how to undertake many common tasks in linear regression (broadly defined), while Chapter 5 discusses many generalizations, including other types of outcome variables, longitudinal and clustered analysis, and survival methods.

Many commands can perform linear regression, as it constitutes a special case of which many models are generalizations. We present detailed descriptions for the `lm()` command, as it offers the most flexibility and best output options tailored to linear regression in particular. While ANOVA can be viewed as a special case of linear regression, separate routines are available (`aov()`) to perform it. We address additional procedures only with respect to output that is difficult to obtain through the standard linear regression tools.

Many of the routines available return or operate on `lm` class objects, which include coefficients, residuals, fitted values, weights, contrasts, model matrices, and the like (see `help(lm)`).

The CRAN Task View on Statistics for the Social Sciences provides an excellent overview of methods described here and in Chapter 5.

4.1 Model fitting

4.1.1 Linear regression

Example: See 4.7.3

```
mod1 = lm(y ~ x1 + ... + xk, data=ds)
summary(mod1)
```

or

```
form = as.formula(y ~ x1 + ... + xk)
mod1 = lm(form, data=ds)
summary(mod1)
```

The first argument of the `lm()` function is a formula object, with the outcome specified followed by the $\sim$ operator then the predictors. More information about the linear model `summary()` command can be found using `help(summary.lm)`. By default, stars are used to annotate the output of the `summary()` functions regarding significance levels: these can be turned off using the command `options(show.signif.stars=FALSE)`.

4.1.2   Linear regression with categorical covariates

Example: See 4.7.3

See also 4.1.3 (parameterization of categorical covariates)

```
x1f = as.factor(x1)
mod1 = lm(y ~ x1f + x2 + ... + xk, data=ds)
```

The `as.factor()` command creates a categorical (or factor/class) variable from a variable. By default, the lowest value (either numerically or by ASCII character code) is the reference value when a factor variable is in a formula. The `levels` option for the `factor()` function can be used to select a particular reference value (see also 2.4.16).

4.1.3   Parameterization of categorical covariates

Example: See 4.7.6

In R, `as.factor()` can be applied before or within any model-fitting function. Parameterization of the covariate can be controlled as in the second example below.

```
mod1 = lm(y ~ as.factor(x))
```

or

```
x.factor = as.factor(x)
mod1 = lm(y ~ x.factor, contrasts=list(x.factor="contr.SAS"))
```

The `as.factor()` function creates a factor object. The `contrasts` option for the `lm()` function specifies how the levels of that factor object should be coded. The `levels` option to the `factor()` function allows specification of the ordering of levels (the default is alphabetical). An example can be found at the beginning of Section 4.7.

The specification of the design matrix for analysis of variance and regression models can be controlled using the `contrasts` option. Examples of options (for a factor with 4 equally spaced levels) are given in Table 4.1. See `options("contrasts")` for defaults, and `contrasts()` or `lm()` to apply a contrast function to a factor variable. Support for reordering factors is available using the `reorder()` function. Ordered factors can be created using the `ordered()` function.

```
> contr.treatment(4)              > contr.poly(4)
  2 3 4                                    .L    .Q      .C
1 0 0 0                           [1,] -0.671   0.5 -0.224
2 1 0 0                           [2,] -0.224  -0.5  0.671
3 0 1 0                           [3,]  0.224  -0.5 -0.671
4 0 0 1                           [4,]  0.671   0.5  0.224
> contr.SAS(4)                    > contr.sum(4)
  1 2 3                               [,1] [,2] [,3]
1 1 0 0                           1     1    0    0
2 0 1 0                           2     0    1    0
3 0 0 1                           3     0    0    1
4 0 0 0                           4    -1   -1   -1
> contr.helmert(4)
   [,1] [,2] [,3]
1   -1   -1   -1
2    1   -1   -1
3    0    2   -1
4    0    0    3
```

Table 4.1: Built-In Options for Contrasts

4.1.4 Linear regression with no intercept

```
mod1 = lm(y ~ 0 + x1 + ... + xk, data=ds)
```
or
```
mod1 = lm(y ~ x1 + ... + xk -1, data=ds)
```

4.1.5 Linear regression with interactions

Example: See 4.7.3
```
mod1 = lm(y ~ x1 + x2 + x1:x2 + x3 + ... + xk, data=ds)
```
or
```
lm(y ~ x1*x2 + x3 + ... + xk, data=ds)
```

The * operator includes all lower order terms, while the : operator includes only the specified interaction. So, for example, the commands y $\sim$ x1*x2*x3 and y $\sim$ x1 + x2 + x3 + x1:x2 + x1:x3 + x2:x3 + x1:x2:x3 have equal values. The syntax also works with any covariates designated as categorical using the `as.factor()` command (see 4.1.2).

4.1.6 Linear models stratified by each value of a grouping variable

Example: See 4.7.5

See also 2.5.1 (subsetting) and 3.1.2 (summary measure by groups)

```
uniquevals = unique(z)
numunique = length(uniquevals)
formula = as.formula(y ~ x1 + ... + xk)
p = length(coef(lm(formula)))
params = matrix(numeric(numunique*p), p, numunique)
for (i in 1:length(uniquevals)) {
    cat(i, "\n")
    params[,i] = coef(lm(formula, subset=(z==uniquevals[i])))
}
```

or

```
modfits = by(ds, z, function(x) lm(y ~ x1 + ... + xk, data=x))
sapply(modfits, coef)
```

In the first codeblock, separate regressions are fit for each value of the grouping variable z through use of a `for` loop. This requires the creation of a matrix of results `params` to be set up in advance, of the appropriate dimension (number of rows equal to the number of parameters (p=k+1) for the model, and number of columns equal to the number of levels for the grouping variable z). Within the loop, the `lm()` function is called and the coefficients from each fit are saved in the appropriate column of the `params` matrix.

The second code block solves the problem using the `by()` function, where the `lm()` function is called for each of the values for z. Additional support for this type of *split-apply-combine* strategy is available in `library(plyr)`.

4.1.7 One-way analysis of variance

Example: See 4.7.6

```
xf = as.factor(x)
mod1 = aov(y ~ xf, data=ds)
summary(mod1)
```

The `summary()` command can be used to provide details of the model fit. More information can be found using `help(summary.aov)`. Note that the function `summary.lm(mod1)` will display the regression parameters underlying the ANOVA model.

4.1.8   Two-way (or more) analysis of variance

See also 4.1.5 (interactions) and 6.1.13 (interaction plots) *Example:* See 4.7.6

```
aov(y ~ as.factor(x1) + as.factor(x2), data=ds)
```

4.2   Model comparison and selection

4.2.1   Compare two models

Example: See 4.7.6

```
mod1 = lm(y ~ x1 + ... + xk, data=ds)
mod2 = lm(y ~ x3 + ... + xk, data=ds)
anova(mod2, mod1)
```
or
```
drop1(mod2)
```

Two nested models may be compared using the `anova()` function. The `anova()` command computes analysis of variance (or deviance) tables. When given one model as an argument, it displays the ANOVA table. When two (or more) nested models are given, it calculates the differences between them. The function `drop1()` computes a table of changes in fit for each term in the named linear model object.

4.2.2   Log-likelihood

See also 4.2.3 (AIC)   *Example:* See 4.7.6

```
mod1 = lm(y ~ x1 + ... + xk, data=ds)
logLik(mod1)
```

The `logLik()` function supports glm, lm, nls, Arima, gls, lme, and nlme objects.

4.2.3   Akaike Information Criterion (AIC)

See also 4.2.2 (log-likelihood)   *Example:* See 4.7.6

```
mod1 = lm(y ~ x1 + ... + xk, data=ds)
AIC(mod1)
```

The `AIC()` function includes support for glm, lm, nls, Arima, gls, lme, and nlme objects. The `stepAIC()` function within `library(MASS)` allows stepwise model selection using AIC (see also 5.4.4, LASSO).

4.2.4   Bayesian Information Criterion (BIC)

See also 4.2.3 (AIC)

```
library(nlme)
mod1 = lm(y ~ x1 + ... + xk, data=ds)
BIC(mod1)
```

4.3   Tests, contrasts, and linear functions of parameters

4.3.1   Joint null hypotheses: Several parameters equal 0

```
mod1 = lm(y ~ x1 + ... + xk, data=ds)
mod2 = lm(y ~ x3 + ... + xk, data=ds)
anova(mod2, mod1)
```

or

```
sumvals = summary(mod1)
covb = vcov(mod1)
coeff.mod1 = coef(mod1)[2:3]
covmat = matrix(c(covb[2,2], covb[2,3],
                  covb[2,3], covb[3,3]), nrow=2)
fval = t(coeff.mod1) %*% solve(covmat) %*% coeff.mod1
pval = 1-pf(fval, 2, mod1$df)
```

The code for the second option, while somewhat complex, builds on the syntax introduced in 4.5.2, 4.5.9, and 4.5.10, and is intended to demonstrate ways to interact with linear model objects.

4.3.2   Joint null hypotheses: Sum of parameters

```
mod1 = lm(y ~ x1 + ... + xk, data=ds)
mod2 = lm(y ~ I(x1+x2-1) + ... + xk, data=ds)
anova(mod2, mod1)
```

or

```
mod1 = lm(y ~ x1 + ... + xk, data=ds)
covb = vcov(mod1)
coeff.mod1 = coef(mod1)
t = (coeff.mod1[2,1]+coeff.mod1[3,1]-1)/
    sqrt(covb[2,2]+covb[3,3]+2*covb[2,3])
pvalue = 2*(1-pt(abs(t), mod1$df))
```

The I() function inhibits the interpretation of operators, to allow them to be used as arithmetic operators. The code in the lower example utilizes the same approach introduced in 4.3.1.

4.3.3 Tests of equality of parameters

Example: See 4.7.8

```
mod1 = lm(y ~ x1 + ... + xk, data=ds)
mod2 = lm(y ~ I(x1+x2) + ... + xk, data=ds)
anova(mod2, mod1)
```

or

```
library(gmodels)
fit.contrast(mod1, "x1", values)
```

or

```
mod1 = lm(y ~ x1 + ... + xk, data=ds)
covb = vcov(mod1)
coeff.mod1 = coef(mod1)
t = (coeff.mod1[2]-coeff.mod1[3])/
    sqrt(covb[2,2]+covb[3,3]-2*covb[2,3])
pvalue = 2*(1-pt(abs(t), mod1$df))
```

The I() function inhibits the interpretation of operators, to allow them to be used as arithmetic operators. The `fit.contrast()` function calculates a contrast in terms of levels of the factor variable `x1` using a numeric matrix vector of contrast coefficients (where each row sums to zero) denoted by `values`. The more general code below utilizes the same approach introduced in 4.3.1 for the specific test of $\beta_1 = \beta_2$ (different coding would be needed for other comparisons).

4.3.4 Multiple comparisons

Example: See 4.7.7

```
mod1 = aov(y ~ x))
TukeyHSD(mod1, "x")
```

The `TukeyHSD()` function takes an argument an `aov` object, and calculates the pairwise comparisons of all of the combinations of the factor levels of the variable `x` (see also `library(multcomp)`).

4.3.5 Linear combinations of parameters

Example: See 4.7.8

It is often useful to calculate predicted values for particular covariate values. Here, we calculate the predicted value $E[Y|X_1 = 1, X_2 = 3] = \hat{\beta}_0 + \hat{\beta}_1 + 3\hat{\beta}_2$.

```
mod1 = lm(y ~ x1 + x2, data=ds)
newdf = data.frame(x1=c(1), x2=c(3))
estimates = predict(mod1, newdf, se.fit=TRUE,
    interval="confidence")
```
or
```
mod1 = lm(y ~ x1 + x2, data=ds)
library(gmodels)
estimable(mod1, c(1, 1, 3))
```

The predict() command can generate estimates at any combination of parameter values, as specified as a dataframe that is passed as an argument. More information on this function can be found using help(predict.lm). Similar functionality is available through the estimable() function.

4.4   Model diagnostics

4.4.1   Predicted values

Example: See 4.7.3

```
mod1 = lm(...)
predicted.varname = predict(mod1)
```

The command predict() operates on any lm() object, and by default generates a vector of predicted values. Similar commands retrieve other regression output.

4.4.2   Residuals

Example: See 4.7.3

```
mod1 = lm(...)
residual.varname = residuals(mod1)
```

The command residuals() operates on any lm() object, and generates a vector of residuals. Other functions for analysis of variance objects, GLM, or linear mixed effects exist (see for example help(residuals.glm)).

4.4.3   Standardized residuals

Example: See 4.7.3

Standardized residuals are calculated by dividing the ordinary residual (observed minus expected, $y_i - \hat{y}_i$) by an estimate of its standard deviation. Studentized residuals are calculated in a similar manner, where the predicted value and the variance of the residual are estimated from the model fit while excluding that observation.

```
mod1 = lm(...)
standardized.resid.varname = stdres(mod1)
studentized.resid.varname = studres(mod1)
```

The `stdres()` and `studres()` functions operate on any `lm()` object, and generate a vector of studentized residuals (the former command includes the observation in the calculation, while the latter does not). Similar commands retrieve other regression output (see `help(influence.measures)`).

4.4.4 Leverage

Example: See 4.7.3

Leverage is defined as the diagonal element of the $(X(X^TX)^{-1}X^T)$ or "hat" matrix.

```
mod1 = lm(...)
leverage.varname = hatvalues(mod1)
```

The command `hatvalues()` operates on any `lm()` object, and generates a vector of leverage values. Similar commands can be utilized to retrieve other regression output (see `help(influence.measures)`).

4.4.5 Cook's D

Example: See 4.7.3

Cook's distance (D) is a function of the leverage (see 4.4.4) and the residual. It is used as a measure of the influence of a data point in a regression model.

```
mod1 = lm(...)
cookd.varname = cooks.distance(mod1)
```

The command `cooks.distance()` operates on any `lm()` object, and generates a vector of Cook's distance values. Similar commands retrieve other regression output.

4.4.6 DFFITS

Example: See 4.7.3

DFFITS are a standardized function of the difference between the predicted value for the observation when it is included in the dataset and when (only) it is excluded from the dataset. They are used as an indicator of the observation's influence.

```
mod1 = lm(...)
dffits.varname = dffits(mod1)
```

The command `dffits()` operates on any `lm()` object, and generates a vector of dffits values. Similar commands retrieve other regression output.

4.4.7   Diagnostic plots

Example: See 4.7.4

```
mod1 = lm(...)
par(mfrow=c(2, 2)) # display 2 x 2 matrix of graphs
plot(mod1)
```

The `plot.lm()` function (which is invoked when `plot()` is given a linear regression model as an argument) can generate six plots: 1) a plot of residuals against fitted values, 2) a Scale-Location plot of $\sqrt{(Y_i - \hat{Y}_i)}$ against fitted values, 3) a normal Q-Q plot of the residuals, 4) a plot of Cook's distances (4.4.5) versus row labels, 5) a plot of residuals against leverages (4.4.4), and 6) a plot of Cook's distances against leverage/(1-leverage). The default is to plot the first three and the fifth. The `which` option can be used to specify a different set (see `help(plot.lm)`).

4.4.8   Heteroscedasticity tests

```
mod1 = lm(y ~ x1 + ... + xk)
library(lmtest)
bptest(y ~ x1 + ... + xk)
```

The `bptest()` function in `library(lmtest)` performs the Breusch-Pagan test for heteroscedasticity [3].

4.5   Model parameters and results

4.5.1   Parameter estimates

Example: See 4.7.3

```
mod1 = lm(...)
coeff.mod1 = coef(mod1)
```

The first element of the vector `coeff.mod1` is the intercept (assuming that a model with an intercept was fit).

4.5.2   Standard errors of parameter estimates

See also 4.5.10 (covariance matrix)

```
mod1 = lm(...)
se.mod1 = coef(summary(mod1))[,2]
```

The standard errors are the second column of the results from `coef()`.

4.5.3 Confidence limits for parameter estimates

Example: See 4.7.3

```
mod1 = lm(...)
confint(mod1)
```

4.5.4 Confidence limits for the mean

Example: See 4.7.2

The lower (and upper) confidence limits for the mean of observations with the given covariate values can be generated, as opposed to the prediction limits for new observations with those values (see 4.5.5).

```
mod1 = lm(...)
pred = predict(mod1, interval="confidence")
lcl.varname = pred[,2]
```

The lower confidence limits are the second column of the results from `predict()`. To generate the upper confidence limits, the user would access the third column of the `predict()` object. The command `predict()` operates on any `lm()` object, and with these options generates confidence limit values. By default, the function uses the estimation dataset, but a separate dataset of values to be used to predict can be specified.

4.5.5 Prediction limits

The lower (and upper) prediction limits for "new" observations can be generated with the covariate values of subjects observed in the dataset (as opposed to confidence limits for the population mean as described in Section 4.5.4).

```
mod1 = lm(...)
pred.w.lowlim = predict(mod1, interval="prediction")[,2]
```

This code saves the second column of the results from the `predict()` function into a vector. To generate the upper confidence limits, the user would access the third column of the `predict()` object. The command `predict()` operates on any `lm()` object, and with these options generates prediction limit values. By default, the function uses the estimation dataset, but a separate dataset of values to be used to predict can be specified.

4.5.6 Plot confidence limits for a particular covariate vector

Example: See 4.7.2

```
pred.w.clim = predict(lm(y ~ x), interval="confidence")
matplot(x, pred.w.clim, lty=c(1, 2, 2), type="l",
   ylab="predicted y")
```

This entry produces fit and confidence limits at the original observations in the original order. If the observations are not sorted relative to the explanatory variable x, the resulting plot will be a jumble. The `matplot()` function is used to generate lines, with a solid line (`lty=1`) for predicted values and dashed line (`lty=2`) for the confidence bounds.

4.5.7   Plot prediction limits for a new observation

Example: See 4.7.2

```
pred.w.plim = predict(lm(y ~ x), interval="prediction")
matplot(x, pred.w.plim, lty=c(1, 2, 2), type="l",
    ylab="predicted y")
```

This entry produces fit and confidence limits at the original observations in the original order. If the observations are not sorted relative to the explanatory variable x, the resulting plot will be a jumble. The `matplot()` function is used to generate lines, with a solid line (`lty=1`) for predicted values and dashed line (`lty=2`) for the confidence bounds.

4.5.8   Plot predicted lines for several values of a predictor

Here we describe how to generate plots for a variable X_1 versus Y separately for each value of the variable X_2 (see also 3.1.2, stratifying by a variable and 6.1.6, conditioning plot).

```
plot(x1, y, pch=" ") # create an empty plot of the correct size
abline(lm(y ~ x1, subset=x2==0), lty=1, lwd=2)
abline(lm(y ~ x1, subset=x2==1), lty=2, lwd=2)
...
abline(lm(y ~ x1, subset=x2==k), lty=k+1, lwd=2)
```

The `abline()` function is used to generate lines for each of the subsets, with a solid line (`lty=1`) for the first group and dashed line (`lty=2`) for the second (this assumes that X_2 takes on values 0–k, see 4.1.6). More sophisticated approaches to this problem can be tackled using `sapply()`, `mapply()`, `split()`, and related functions.

4.5.9   Design and information matrix

See also 2.9 (matrices) and 4.1.3 (parametrization of design matrices).

```
mod1 = lm(y ~ x1 + ... + xk, data=ds)
XpX = t(model.matrix(mod1)) %*% model.matrix(mod1)
```

or

```
X = cbind(rep(1, length(x1)), x1, x2, ..., xk)
XpX = t(X) %*% X
rm(X)
```

The `model.matrix()` function creates the design matrix from a linear model object. Alternatively, this quantity can be built up using the `cbind()` function to glue together the design matrix X. Finally, matrix multiplication and the transpose function are used to create the information $(X'X)$ matrix.

4.5.10   Covariance matrix of the predictors

See also 2.9 (matrices) and 4.5.2 (standard errors)   *Example:* See 4.7.3

```
mod1 = lm(...)
varcov = vcov(mod1)
```
or
```
sumvals = summary(mod1)
covb = sumvals$cov.unscaled*sumvals$sigma^2
```

Running `help(summary.lm)` provides details on return values.

4.6   Further resources

Faraway [14] provides an accessible guide to linear regression in R, while Cook [7] details a variety of regression diagnostics. The CRAN Task View on Statistics for the Social Sciences provides an excellent overview of methods described here and in Chapter 5.

4.7   HELP examples

To help illustrate the tools presented in this chapter, we apply many of the entries to the HELP data. The code for these examples can be downloaded from http://www.math.smith.edu/r/examples.

We begin by reading in the dataset and keeping only the female subjects. We create a version of the `substance` variable as a factor (see 4.1.3).

```
> options(digits=3)
> options(width=67) # narrow output
> library(foreign)
> ds = read.csv("http://www.math.smith.edu/r/data/help.csv")
> newds = ds[ds$female==1,]
> attach(newds)
> sub = factor(substance, levels=c("heroin", "alcohol",
+    "cocaine"))
```

4.7.1   Scatterplot with smooth fit

As a first step to help guide fitting a linear regression, we create a scatterplot (6.1.1) displaying the relationship between age and the number of alcoholic drinks consumed in the period before entering detox (variable name: i1), as well as primary substance of abuse (alcohol, cocaine, or heroin).

Figure 4.1 displays a scatterplot of observed values for i1 (along with separate smooth fits by primary substance). To improve legibility, the plotting region is restricted to those with number of drinks between 0 and 40 (see plotting limits, 6.3.7).

```
> plot(age, i1, ylim=c(0,40), type="n", cex.lab=1.4,
+    cex.axis=1.4)
> points(age[substance=="alcohol"], i1[substance=="alcohol"],
+    pch="a")
> lines(lowess(age[substance=="alcohol"],
+    i1[substance=="alcohol"]), lty=1, lwd=2)
> points(age[substance=="cocaine"], i1[substance=="cocaine"],
+    pch="c")
> lines(lowess(age[substance=="cocaine"],
+    i1[substance=="cocaine"]), lty=2, lwd=2)
> points(age[substance=="heroin"], i1[substance=="heroin"],
+    pch="h")
> lines(lowess(age[substance=="heroin"],
+    i1[substance=="heroin"]), lty=3, lwd=2)
> legend(44, 38, legend=c("alcohol", "cocaine", "heroin"),
+    lty=1:3, cex=1.4, lwd=2, pch=c("a", "c", "h"))
```

The pch option to the legend() command can be used to insert plot symbols in legends (Figure 4.1 displays the different line styles).

Not surprisingly, Figure 4.1 suggests that there is a dramatic effect of primary substance, with alcohol users drinking more than others. There is some indication of an interaction with age.

4.7.2   Regression with prediction intervals

We demonstrate plotting confidence limits (4.5.4) as well as prediction limits (4.5.7) from a linear regression model of pcs as a function of age.

We first sort the data, as needed by matplot(). Figure 4.2 displays the predicted line along with these intervals.

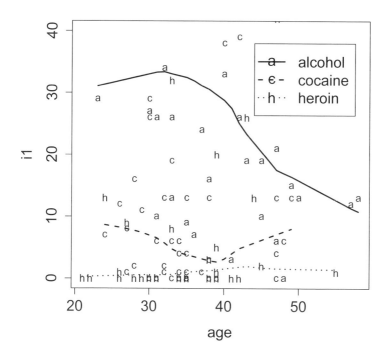

Figure 4.1: Scatterplot of observed values for AGE and I1 (plus smoothers by substance).

```
> ord = order(age)
> orderage = age[ord]
> orderpcs = pcs[ord]
> lm1 = lm(orderpcs ~ orderage)
> pred.w.clim = predict(lm1, interval="confidence")
> pred.w.plim = predict(lm1, interval="prediction")
> matplot(orderage, pred.w.plim, lty=c(1, 2, 2), type="l",
+     ylab="predicted PCS", xlab="age (in years)", lwd=2)
> matpoints(orderage, pred.w.clim, lty=c(1, 3, 3), type="l",
+     lwd=2)
> legend(40, 56, legend=c("prediction", "confidence"), lty=2:3,
+     lwd=2)
```

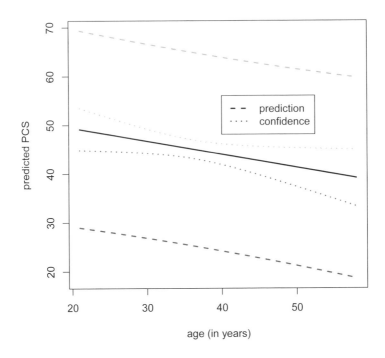

Figure 4.2: Predicted values for PCS as a function of age (plus confidence and prediction intervals).

4.7.3   Linear regression with interaction

Next we fit a linear regression model (4.1.1) for the number of drinks as a function of age, substance, and their interaction (4.1.5). To assess the need for the interaction, we fit the model with no interaction and use the `anova()` function to compare the models (the `drop1()` function could also be used).

```
> options(show.signif.stars=FALSE)
> lm1 = lm(i1 ~ sub * age)
> lm2 = lm(i1 ~ sub + age)
> anova(lm2, lm1)

Analysis of Variance Table

Model 1: i1 ~ sub + age
Model 2: i1 ~ sub * age
  Res.Df   RSS Df Sum of Sq    F Pr(>F)
1    103 26196
2    101 24815  2      1381 2.81  0.065
```

There is some indication of a borderline significant interaction between age and substance group (p=0.065).

There are many quantities of interest stored in the linear model object lm1, and these can be viewed or extracted for further use.

```
> names(summary(lm1))

 [1] "call"          "terms"         "residuals"
 [4] "coefficients"  "aliased"       "sigma"
 [7] "df"            "r.squared"     "adj.r.squared"
[10] "fstatistic"    "cov.unscaled"

> summary(lm1)$sigma

[1] 15.7

> names(lm1)

 [1] "coefficients"  "residuals"      "effects"
 [4] "rank"          "fitted.values" "assign"
 [7] "qr"            "df.residual"   "contrasts"
[10] "xlevels"       "call"          "terms"
[13] "model"
```

```
> lm1$coefficients

   (Intercept)      subalcohol       subcocaine            age
        -7.770          64.880           13.027          0.393
subalcohol:age subcocaine:age
        -1.113          -0.278

> coef(lm1)

   (Intercept)      subalcohol       subcocaine            age
        -7.770          64.880           13.027          0.393
subalcohol:age subcocaine:age
        -1.113          -0.278

> confint(lm1)

                   2.5 %   97.5 %
(Intercept)      -33.319   17.778
subalcohol        28.207  101.554
subcocaine       -24.938   50.993
age               -0.325    1.112
subalcohol:age    -2.088   -0.138
subcocaine:age    -1.348    0.793

> vcov(lm1)

                (Intercept) subalcohol subcocaine      age
(Intercept)          165.86    -165.86    -165.86   -4.548
subalcohol          -165.86     341.78     165.86    4.548
subcocaine          -165.86     165.86     366.28    4.548
age                   -4.55       4.55       4.55    0.131
subalcohol:age         4.55      -8.87      -4.55   -0.131
subcocaine:age         4.55      -4.55     -10.13   -0.131
                subalcohol:age subcocaine:age
(Intercept)              4.548          4.548
subalcohol              -8.866         -4.548
subcocaine              -4.548        -10.127
age                     -0.131         -0.131
subalcohol:age           0.241          0.131
subcocaine:age           0.131          0.291
```

4.7.4 Regression diagnostics

Assessing the model is an important part of any analysis. We begin by examining the residuals (4.4.2). First, we calculate the quantiles of their distribution, then display the smallest residual.

```
> pred = fitted(lm1)
> resid = residuals(lm1)
> quantile(resid)

    0%     25%     50%     75%    100%
-31.92   -8.25   -4.18    3.58   49.88
```

We could examine the output, then select a subset of the dataset to find the value of the residual that is less than −31. Instead the dataset can be sorted so the smallest observation is first and then print the minimum observation.

```
> tmpds = data.frame(id, age, i1, sub, pred, resid,
+     rstandard(lm1))
> tmpds[resid==max(resid),]

   id age i1      sub pred resid rstandard.lm1.
4   9  50 71 alcohol 21.1  49.9           3.32

> tmpds[resid==min(resid),]

    id age i1      sub pred resid rstandard.lm1.
72 325  35  0 alcohol 31.9 -31.9          -2.07
```

The output includes the row number of the minimum and maximum residual.

Graphical tools are the best way to examine residuals. Figure 4.3 displays the default diagnostic plots (4.4) from the model.

```
> oldpar = par(mfrow=c(2, 2), mar=c(4, 4, 2, 2)+.1)
> plot(lm1)
> par(oldpar)
```

Figure 4.4 displays the empirical density of the standardized residuals, along with an overlaid normal density. The assumption that the residuals are approximately Gaussian does not appear to be tenable.

```
> library(MASS)
> std.res = rstandard(lm1)
> hist(std.res, breaks=seq(-2.5, 3.5, by=.5), main="",
+     xlab="standardized residuals", col="gray80", freq=FALSE)
> lines(density(std.res), lwd=2)
> xvals = seq(from=min(std.res), to=max(std.res), length=100)
> lines(xvals, dnorm(xvals, mean(std.res), sd(std.res)), lty=2,
+     lwd=3)
```

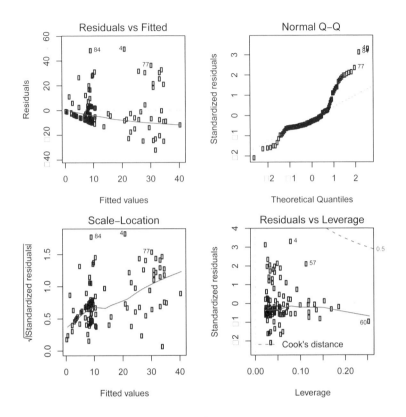

Figure 4.3: Default diagnostics.

The residual plots indicate some potentially important departures from model assumptions, and further exploration should be undertaken.

4.7.5   Fitting regression model separately for each value of another variable

One common task is to perform identical analyses in several groups. Here, as an example, we consider separate linear regressions for each substance abuse group.

A matrix of the correct size is created, then a `for` loop is run for each unique value of the grouping variable.

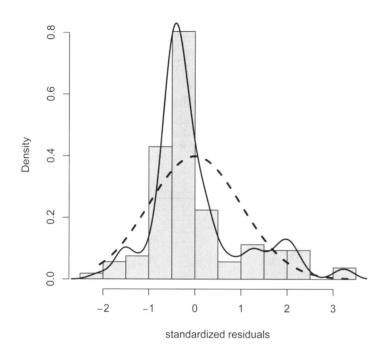

Figure 4.4: Empirical density of residuals, with superimposed normal density.

```
> uniquevals = unique(substance)
> numunique = length(uniquevals)
> formula = as.formula(i1 ~ age)
> p = length(coef(lm(formula)))
> res = matrix(rep(0, numunique*p), p, numunique)
> for (i in 1:length(uniquevals)) {
+    res[,i] = coef(lm(formula, subset=substance==uniquevals[i]))
+ }
> rownames(res) = c("intercept","slope")
> colnames(res) = uniquevals
> res

          heroin cocaine alcohol
intercept -7.770   5.257   57.11
slope      0.393   0.116   -0.72

> detach(newds)
```

4.7.6   Two-way ANOVA

Is there a statistically significant association between gender and substance abuse group with depressive symptoms? The function `interaction.plot()` can be used to graphically assess this question. Figure 4.5 displays an interaction plot for CESD as a function of substance group and gender.

```
> attach(ds)
> sub = as.factor(substance)
> gender = as.factor(ifelse(female, "F", "M"))
> interaction.plot(sub, gender, cesd, xlab="substance", las=1,
+     lwd=2)
```

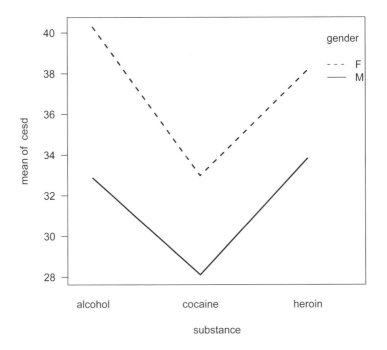

Figure 4.5: Interaction plot of CESD as a function of substance group and gender.

There are indications of large effects of gender and substance group, but little suggestion of interaction between the two. The same conclusion is reached in Figure 4.6, which displays boxplots by substance group and gender.

```
> subs = character(length(substance))
> subs[substance=="alcohol"] = "Alc"
> subs[substance=="cocaine"] = "Coc"
> subs[substance=="heroin"] = "Her"
> gen = character(length(female))
> boxout = boxplot(cesd ~ subs + gender, notch=TRUE,
+     varwidth=TRUE, col="gray80")
> boxmeans = tapply(cesd, list(subs, gender), mean)
> points(seq(boxout$n), boxmeans, pch=4, cex=2)
```

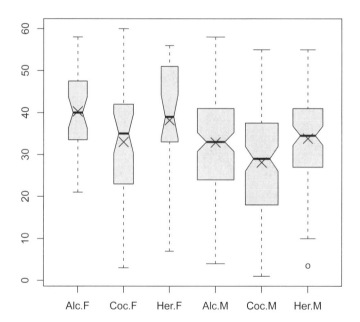

Figure 4.6: Boxplot of CESD as a function of substance group and gender.

The width of each box is proportional to the size of the sample, with the notches denoting confidence intervals for the medians, and X's marking the observed means.

Next, we proceed to formally test whether there is a significant interaction through a two-way analysis of variance (4.1.8). We fit models with and without an interaction, and then compare the results. We also construct the likelihood ratio test manually.

```
> aov1 = aov(cesd ~ sub * gender, data=ds)
> aov2 = aov(cesd ~ sub + gender, data=ds)
> resid = residuals(aov2)
> anova(aov2, aov1)

Analysis of Variance Table

Model 1: cesd ~ sub + gender
Model 2: cesd ~ sub * gender
  Res.Df    RSS Df Sum of Sq    F Pr(>F)
1    449 65515
2    447 65369  2       146 0.5   0.61

> options(digits=6)
> logLik(aov1)

'log Lik.' -1768.92 (df=7)

> logLik(aov2)

'log Lik.' -1769.42 (df=5)

> lldiff = logLik(aov1)[1] - logLik(aov2)[1]
> lldiff

[1] 0.505055

> 1 - pchisq(2*lldiff, 2)

[1] 0.603472

> options(digits=3)
```

There is little evidence (p=0.61) of an interaction, so this term can be dropped.
The model was previously fit to test the interaction, and can be displayed.

```
> aov2

Call:
   aov(formula = cesd ~ sub + gender, data = ds)

Terms:
                   sub gender Residuals
Sum of Squares    2704   2569     65515
Deg. of Freedom      2      1       449

Residual standard error: 12.1
Estimated effects may be unbalanced

> summary(aov2)

             Df Sum Sq Mean Sq F value  Pr(>F)
sub           2   2704    1352    9.27 0.00011
gender        1   2569    2569   17.61 3.3e-05
Residuals   449  65515     146
```

The default design matrix (lowest value is reference group, see 4.1.3) can be changed and the model refit. In this example, we specify the coding where the highest value is denoted as the reference group (which could allow matching results from a similar model fit in SAS).

```
> contrasts(sub) = contr.SAS(3)
> aov3 = lm(cesd ~ sub + gender, data=ds)
> summary(aov3)

Call:
lm(formula = cesd ~ sub + gender, data = ds)

Residuals:
    Min      1Q Median     3Q    Max
 -32.13   -8.85   1.09   8.48  27.09

Coefficients:
             Estimate Std. Error t value Pr(>|t|)
(Intercept)    39.131      1.486   26.34  < 2e-16
sub1           -0.281      1.416   -0.20  0.84247
sub2           -5.606      1.462   -3.83  0.00014
genderM        -5.619      1.339   -4.20  3.3e-05

Residual standard error: 12.1 on 449 degrees of freedom
Multiple R-squared: 0.0745,        Adjusted R-squared: 0.0683
F-statistic:   12 on 3 and 449 DF,  p-value: 1.35e-07
```

The AIC criteria (4.2.3) can also be used to compare models: this also suggests that the model without the interaction is most appropriate.

```
> AIC(aov1)

[1] 3552

> AIC(aov2)

[1] 3549
```

4.7.7 Multiple comparisons

We can also carry out multiple comparison (4.3.4) procedures to test each of the pairwise differences between substance abuse groups. We use the `TukeyHSD()` function here.

```
> mult = TukeyHSD(aov(cesd ~ sub, data=ds), "sub")
> mult

  Tukey multiple comparisons of means
    95% family-wise confidence level

Fit: aov(formula = cesd ~ sub, data = ds)

$sub
                  diff    lwr   upr p adj
cocaine-alcohol -4.952 -8.15 -1.75 0.001
heroin-alcohol   0.498 -2.89  3.89 0.936
heroin-cocaine   5.450  1.95  8.95 0.001
```

The alcohol group and heroin group both have significantly higher CESD scores than the cocaine group, but the alcohol and heroin groups do not significantly differ from each other (95% CI ranges from −2.8 to 3.8). Figure 4.7 provides a graphical display of the pairwise comparisons.

```
> plot(mult)
```

4.7.8 Contrasts

We can also fit contrasts (4.3.3) to test hypotheses involving multiple parameters. In this case, we can compare the CESD scores for the alcohol and heroin groups to the cocaine group.

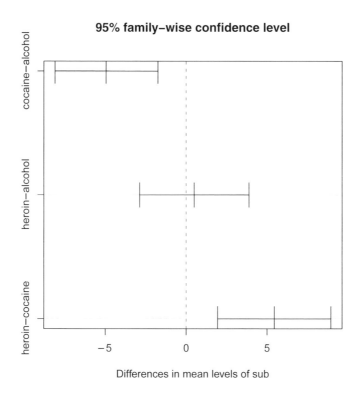

Figure 4.7: Pairwise comparisons.

```
> library(gmodels)
> fit.contrast(aov2, "sub", c(1,-2,1), conf.int=0.95 )

                 Estimate Std. Error t value Pr(>|t|) lower CI
sub c=( 1 -2 1 )     10.9       2.42    4.52 8.04e-06     6.17
                 upper CI
sub c=( 1 -2 1 )     15.7
```

As expected from the interaction plot (Figure 4.5), there is a statistically significant difference in this one degree of freedom comparison (p<0.0001).

Chapter 5

Regression generalizations and multivariate statistics

This chapter discusses many commonly used statistical models beyond linear regression and ANOVA, as well as multivariate statistics. The CRAN Task Views on statistics for the social sciences, psychometric models, and multivariate statistics provide useful overviews.

5.1 Generalized linear models

Table 5.1 displays the options to specify link functions and family of distributions for generalized linear models [43]. Description of several specific generalized linear regression models (e.g., logistic and Poisson) can be found in subsequent sections of this chapter.

```
glm(y ~ x1 + ... + xk, family="familyname"(link="linkname"),
    data=ds)
```

More information on GLM families and links can be found using `help(family)`.

5.1.1 Logistic regression model

Example: See 5.7.1 and 7.7

```
glm(y ~ x1 + ... + xk, binomial, data=ds)
```

or

```
library(rms)
lrm(y ~ x1 + ... + xk, data=ds)
```

Distribution	Options for command `glm()`
Gaussian	`family="gaussian"`, `link="identity"`, `"log"` or `"inverse"`
binomial	`family="binomial"`, `link="logit"`, `"probit"`, `"cauchit"`, `"log"` or `"cloglog"`
gamma	`family="Gamma"`, `link="inverse"`, `"identity"` or `"log"`
Poisson	`family="poisson"`, `link="log"`, `"identity"` or `"sqrt"`
inverse Gaussian	`family="inverse.gaussian"`, `link="1/mu^2"`, `"inverse"`, `"identity"` or `"sqrt"`
Multinomial	See `multinom()` in `nnet` library
Negative Binomial	See `negative.binomial()` in `MASS` library
overdispersed	`family="quasi"`, `link="logit"`, `"probit"`, `"cloglog"`, `"identity"`, `"inverse"`, `"log"`, `"1/mu^2"` or `"sqrt"` (see also `glm.binomial.disp()` in the `dispmod` library)

Table 5.1: Generalized Linear Model Distributions

The `lrm()` function within the `rms` package provides the so-called "c" statistic (area under ROC curve, see also 6.1.17) and the Nagelkerke pseudo-R^2 index [47].

5.1.2   Exact logistic regression

```
library(elrm)
elrmres = elrm(y ~ x1 + ... + xk, iter=10000, burnIn=10000,
    data=ds)
```

The `elrm()` function implements a modified MCMC algorithm to approximate exact conditional inference for logistic regression models [80] (see also 5.4.7, Bayesian methods).

5.1.3   Poisson model

See also 5.1.5 (zero-inflated Poisson)   *Example:* See 5.7.2

```
glm(y ~ x1 + ... + xk, poisson, data=ds)
```

5.1.4 Goodness of fit for count models

It is always important to check assumptions for models. This is particularly true for Poisson models, which are quite sensitive to model departures [26]. One way to assess the fit of the model is by comparing the observed and expected cell counts, and then calculating Pearson's chi-square statistic. This can be carried out using the goodfit() function.

```
library(vcd)
poisfit = goodfit(x, "poisson")
```

The goodfit() function carries out a Pearson's χ^2 test of observed vs. expected counts. Other distributions supported include binomial and nbinomial. A hanging rootogram [74] can also be generated to assess the goodness of fit for count models. If the model fits well, then the bottom of each bar in the rootogram should be near zero.

```
library(vcd)
rootogram(poisfit)
```

5.1.5 Zero-inflated Poisson model

Example: See 5.7.3

Zero-inflated Poisson models can be used for count outcomes that generally follow a Poisson distribution but for which there are (many) more observed counts of 0 than would be expected. These data can be seen as deriving from a mixture distribution of a Poisson and a degenerate distribution with point mass at zero (see also 5.1.7, zero-inflated negative binomial).

```
library(pscl)
mod = zeroinfl(y ~ x1 + ... + xk | x2 + ... + xp, data=ds)
```

The Poisson rate parameter of the model is specified in the usual way with a formula as argument to zeroinfl(). The default link is log. The excess zero-probability is modeled as a function of the covariates specified after the "|" character. An intercept-only model can be fit by including 1 as the second model. Zero-inflated negative binomial and geometric models are also supported.

5.1.6 Negative binomial model

See also 5.1.7 (zero-inflated negative binomial) *Example:* See 5.7.4

```
library(MASS)
glm.nb(y ~ x1 + ... + xk, data=ds)
```

5.1.7   Zero-inflated negative binomial model

Zero-inflated negative binomial models can be used for count outcomes that generally follow a negative binomial distribution but for which there are (many) more observed counts of 0 than would be expected. These data can be seen as deriving from a mixture distribution of a negative binomial and a degenerate distribution with point mass at zero (see also 5.1.5, zero-inflated Poisson).

```
library(pscl)
mod = zeroinfl(y ~ x1 + ... + xk | x2 + ... + xp, dist="negbin",
   data=ds)
```

The negative binomial rate parameter of the model is specified in the usual way with a formula as argument to `zeroinfl()`. The default link is `log`. The zero-probability is modeled as a function of the covariates specified after the "|" character. An intercept-only model can be fit by including `1` as the model.

5.1.8   Log-linear model

Loglinear models are a flexible approach to analysis of categorical data [1]. A loglinear model of a three-dimensional contingency table denoted by X_1, X_2, and X_3 might assert that the expected counts depend on a two-way interaction between the first two variables, but that X_3 is independent of all the others:

$$log(m_{ijk}) = \mu + \lambda_i^{X_1} + \lambda_j^{X_2} + \lambda_{ij}^{X_1,X_2} + \lambda_k^{X_3}$$

```
logres = loglin(table(x1, x2, x3), margin=list(c(1,2), c(3)),
   param=TRUE)
pvalue = 1-pchisq(logres$lrt, logres$df)
```

The `margin` option specifies the dependence assumptions. In addition to the `loglin()` function, the `loglm()` function within the `MASS` library provides an interface for log-linear modeling.

5.1.9   Ordered multinomial model

Example: See 5.7.7

```
library(MASS)
polr(y ~ x1 + ... + xk, data=ds)
```

The default link is logistic; this can be changed to probit, complementary log-log or Cauchy using the `method` option.

5.1.10   Generalized (nominal outcome) multinomial logit

Example: See 5.7.8

```
library(VGAM)
mlogit = vglm(y ~ x1 + ... + xk, family=multinomial(), data=ds)
```

5.1.11   Conditional logistic regression model

```
library(survival)
cmod = clogit(y ~ x1 + ... + xk + strata(id), data=ds)
```

The variable `id` identifies strata or matched sets of observations. An exact model is fit by default.

5.2   Models for correlated data

There is extensive support within for correlated data regression models, including repeated measures, longitudinal, time series, clustered, and other related methods. Throughout this section we assume that repeated measurements are taken on a subject or cluster denoted by variable `id`.

5.2.1   Linear models with correlated outcomes

Example: See 5.7.11

```
library(nlme)
glsres = gls(y ~ x1 + ... + xk,
    correlation=corSymm(form = ~ ordervar | id),
    weights=varIdent(form = ~1 | ordervar), ds)
```

The `gls()` function supports estimation of generalized least squares regression models with arbitrary specification of the variance covariance matrix. In addition to a formula interface for the mean model, the analyst specifies a within-group correlation structure as well as a description of the within-group heteroscedasticity structure (using the `weights` option). The statement `ordervar | id` implies that associations are assumed within `id`. Other covariance matrix options are available, see `help(corClasses)`.

5.2.2   Linear mixed models with random intercepts

See also 5.2.3 (random slope models), 5.2.4 (random coefficient models), and 7.1.2 (empirical power calculations)

```
library(nlme)
lmeint = lme(fixed= y ~ x1 + ... + xk, random = ~ 1 | id,
    na.action=na.omit, data=ds)
```

Best linear unbiased predictors (BLUPs) of the sum of the fixed effects plus corresponding random effects can be generated using the `coef()` function, random effect estimates using the `random.effects()` function, and the estimated variance covariance matrix of the random effects using `VarCorr()`. The fixed effects can be returned using the `fixed.effects()` command. Normalized residuals (using a Cholesky decomposition, see Fitzmaurice, Laird, and Ware [17]) can be generated using the `type="normalized"` option when calling `residuals()` (more information can be found using `help(residuals.lme)`). A plot of the random effects can be created using `plot(lmeint)`.

5.2.3   Linear mixed models with random slopes

Example: See 5.7.12

See also 5.2.2 (random intercept models) and 5.2.4 (random coefficient models)

```
library(nlme)
lmeslope = lme(fixed=y ~ time + x1 + ... + xk,
    random = ~ time | id, na.action=na.omit, data=ds)
```

The default covariance for the random effects is unstructured (other options are described in `help(reStruct)`). Best linear unbiased predictors (BLUPs) of the sum of the fixed effects plus corresponding random effects can be generated using the `coef()` function, random effect estimates using the `random.effects()` function, and the estimated variance covariance matrix of the random effects using `VarCorr()`. A plot of the random effects can be created using `plot(lmeint)`.

5.2.4   More complex random coefficient models

We can extend the random effects models introduced in 5.2.2 and 5.2.3 to 3 or more subject-specific random parameters (e.g., a quadratic growth curve or spline/"broken stick" model [17]). We use `time1` and `time2` to refer to 2 generic functions of time.

```
library(nlme)
lmestick = lme(fixed= y ~ time1 + time2 + x1 + ... + xk,
    random = ~ time1 time2 | id, data=ds, na.action=na.omit)
```

The default covariance for the random effects is unstructured (other options are described in `help(reStruct)`). Best linear unbiased predictors (BLUPs) of the sum of the fixed effects plus corresponding random effects can be generated using the `coef()` function, random effect estimates using the `random.effects()` function, and the estimated variance covariance matrix of the random effects using `VarCorr()`. A plot of the random effects can be created using `plot(lmeint)`.

5.2.5   Multilevel models

Studies with multiple levels of clustering can be fit. In a typical example, a study might include schools (as one level of clustering) and classes within schools (a second level of clustering), with individual students within the classrooms providing a response. Generically, we refer to $level_l$ variables which are identifiers of cluster membership at level l. Random effects at different levels are assumed to be uncorrelated with each other.

```
library(nlme)
lmres = lme(fixed= y ~ x1 + ... + xk,
   random= ~ 1 | level1 / level2, data=ds)
```

A model with k levels of clustering can be fit using the syntax: `level1 / ... / levelk`.

5.2.6   Generalized linear mixed models

Example: See 5.7.14 and 7.1.2

```
library(lme4)
glmmres = lmer(y ~ x1 + ... + xk + (1|id), family=familyval,
   data=ds)
```

See `help(family)` for details regarding specification of distribution families and link functions.

5.2.7   Generalized estimating equations

Example: See 5.7.13

```
library(gee)
geeres = gee(formula = y ~ x1 + ... + xk, id=id, data=ds,
   family=binomial, corstr="independence")
```

The `gee()` function requires that the dataframe be sorted by subject identifier (see 2.5.6). Other correlation structures include `"AR-M"`, `"fixed"`, `"stat_M_dep"`, `"non_stat_M_dep"`, and `"unstructured"`. Note that the `"unstructured"` working correlation will only yield correct answers when missing data are monotone, since no ordering options are available in the present release (see `help(gee)` for more information).

5.2.8   Time-series model

Time-series modeling is an extensive area with a specialized language and notation. We make only the briefest approach here. We display fitting an ARIMA (autoregressive integrated moving average) model for the first difference, with first-order auto-regression and moving averages. The CRAN Task View on Time Series provides an overview of relevant routines.

```
tsobj = ts(x, frequency=12, start=c(1992, 2))
arres = arima(tsobj, order=c(1, 1, 1))
```

The `ts()` function creates a time series object, in this case for monthly time-series data within the variable x beginning in February 1992 (the default behavior is that the series starts at time 1 and number of observations per unit of time is 1). The `start` option is either a single number or a vector of two integers which specify a natural time unit and a number of samples into the time unit. The `arima()` function fits an ARIMA model with AR, differencing and MA order all equal to 1.

5.3   Survival analysis

Survival (or failure time) data consist of the time until an event is observed, as well as an indicator of whether the event was observed or censored at that time. Throughout, we denote the time of measurement with the variable `time` and censoring with a dichotomous variable `cens` $= 1$ if censored, or $= 0$ if observed.   More information on survival (or failure time, or time-to-event) analysis can be found in the CRAN Survival Analysis Task View (see 1.7.2). Other entries related to survival analysis include 3.4.4 (logrank test) and 6.1.18 (Kaplan–Meier plot).

5.3.1   Proportional hazards (Cox) regression model

Example: See 5.7.15 and 7.2.4

```
library(survival)
coxph(Surv(time, cens) ~ x1 + ... + xk)
```

The Efron estimator is the default; other choices including exact and Breslow can be specified using the `method` option. The `cph()` function within the `rms` package supports time varying covariates, while the `cox.zph()` function within the `survival` package allows testing of the proportionality assumption. See `survfit()` for estimates of the the baseline cumulative hazard and other related quantities.

5.3.2   Proportional hazards (Cox) model with frailty

```
library(survival)
coxph(Surv(time, cens) ~ x1 + ... + xk + frailty(id), data=ds)
```

More information on specification of frailty models can be found using the command `help(frailty)`; support is available for t, Gamma and Gaussian distributions. Additional functionality to fit frailty models using maximum penalized likelihood estimation is available in `library(frailtypack)`.

5.4 Further generalizations to regression models

5.4.1 Nonlinear least squares model

Nonlinear least squares models [64] can be flexibly fit. As an example, consider the income inequality model described by Sarabia and colleagues [60]:

$$Y = (1 - (1 - X)^p)^{(1/p)}$$

```
nls(y ~ (1- (1-x)^{p})^(1/{p}), start=list(p=0.5), trace=TRUE)
```

We provide a starting value (0.5) within the interior of the parameter space. Finding solutions for nonlinear least squares problems is often challenging, see `help(nls)` for information on supported algorithms as well as Section 2.8.8 (optimization).

5.4.2 Generalized additive model

Example: See 5.7.9

```
library(gam)
gam(y ~ s(x1, df) + lo(x2) + lo(x3, x4) + x5 + ... + xk, data=ds)
```

Specification of a smooth term for variable `x1` is given by `s(x1)`, while a univariate or bivariate loess fit can be included using `lo(x1, x2)`. See `gam.s()` and `gam.lo()` within `library(gam)` for details regarding specification of degrees of freedom or span, respectively. Polynomial regression terms can be fit using the `poly()` function. Support for additive mixed models is provided by `library(amer)`.

5.4.3 Robust regression model

Robust regression refers to methods for detecting outliers and/or providing stable estimates when they are present. Outlying variables in the outcome, predictor, or both are considered.

```
library(MASS)
rlm(y ~ x1 + ... + xk, data=ds)
```

The `rlm()` function fits a robust linear model using M estimation. More information can be found in the CRAN Robust Statistical Methods Task View.

5.4.4　LASSO model selection

Example: See 5.7.5

The LASSO (least absolute shrinkage and selection operator) is a model selection method for linear regression that minimizes the sum of squared errors subject to a constraint on the sum of the absolute value of the coefficients [69]. This is particularly useful in data mining situations where a large number of predictors are being considered for inclusion in the model.

```
library(lars)
lars(y ~ x1 + ... + xk, data=ds, type="lasso")
```

The `lars()` function also implements least angle regression and forward stagewise methods.

5.4.5　Quantile regression model

Example: See 5.7.6

Quantile regression predicts changes in the specified quantile of the outcome variable per unit change in the predictor variables; analogous to the change in the mean predicted in least squares regression. If the quantile so predicted is the median, this is equivalent to minimum absolute deviation regression (as compared to least squares regression minimizing the squared deviations).

```
library(quantreg)
quantmod = rq(y ~ x1 + ... + xk, tau=0.75, data=ds)
```

The default for `tau` is 0.5, corresponding to median regression. If a vector is specified, the return value includes a matrix of results.

5.4.6　Ridge regression model

Ridge regression is an extension of multiple regression when predictors are nearly collinear.

```
library(MASS)
ridgemod = lm.ridge(y ~ x1 + ... + xk,
    lambda=seq(from=a, to=b, by=c), data=ds)
```

Postestimation functions supporting `lm.ridge()` objects include `plot()` and `select()`. A vector of ridge constants can be specified using the `lambda` option.

5.4.7 Bayesian methods

Example: See 5.7.16

Bayesian methods are increasingly commonly utilized, and implementations of many models are available. The CRAN Bayesian Inference Task View provides an overview of the packages that incorporate some aspect of Bayesian methodologies.

```
library(MCMCpack)

# linear regression
mod1 = MCMCregress(formula, burnin=1000, mcmc=10000, data=ds)

# logistic regression
mod2 = MCMClogit(formula, burnin=1000, mcmc=10000, data=ds)

# Poisson regression
mod3 = MCMCpoisson(formula, burnin=1000, mcmc=10000, data=ds)
```

Table 5.2 displays modeling functions available within the `MCMCpack` library (including the three listed above). The default prior means are set to zero with precision given by an improper uniform prior.

Specification of prior distributions is important for Bayesian analysis. In addition, diagnosis of convergence is a critical part of any MCMC model fitting (see Gelman et al. [21] for an accessible introduction). Support for model assessment is provided in the `coda` (Convergence Diagnosis and Output Analysis) package, which can operate on `mcmc` objects returned by the `MCMC` routines.

5.4.8 Complex survey design

The appropriate analysis of sample surveys requires incorporation of complex design features, including stratification, clustering, weights, and finite population correction. These design components can be incorporated for many common models. In this example, we assume that there are variables `psuvar` (cluster or PSU), `stratum` (stratification variable), and `wt` (sampling weight). Code examples are given to estimate the mean of a variable `x1` as well as a linear regression model.

```
library(survey)
mydesign = svydesign(id=~psuvar, strata=~stratum, weights=~wt,
    fpc=~fpcvar, data=ds)
meanres = svymean(~ x1, mydesign)
regres = svyglm(y ~ x1 + ... + xk, design=mydesign)
```

The `survey` library includes support for many models. Illustrated above are means and linear regression models, with specification of PSU's, stratification, weight, and FPC [8].

MCMCSVDreg()	MCMC for SVD Regression
MCMCbinaryChange()	MCMC for a Binary Multiple Changepoint Model
MCMCdynamicEI()	MCMC for Quinn's Dynamic Ecological Inference Model
MCMCdynamicIRT1d()	MCMC for Dynamic One-Dimensional Item Response Theory Model
MCMCfactanal()	MCMC for Normal Theory Factor Analysis Model
MCMChierEI()	MCMC for Wakefield's Hierarchical Ecological Inference Model
MCMCirt1d()	MCMC for One-Dimensional Item Response Theory Model
MCMCirtHier1d()	MCMC for Hierarchical One-Dimensional Item Response Theory Model, Covariates Predicting Latent Ideal Point (Ability)
MCMCirtKd()	MCMC for K-Dimensional Item Response Theory Model
MCMCirtKdHet()	MCMC for Heteroskedastic K-Dimensional Item Response Theory Model
MCMCirtKdRob()	MCMC for Robust K-Dimensional Item Response Theory Model
MCMClogit()	MCMC for Logistic Regression
MCMCmetrop1R()	Metropolis Sampling from User-Written R function
MCMCmixfactanal()	MCMC for Mixed Data Factor Analysis Model
MCMCmnl()	MCMC for Multinomial Logistic Regression
MCMCoprobit()	MCMC for Ordered Probit Regression
MCMCordfactanal()	MCMC for Ordinal Data Factor Analysis Model
MCMCpoisson()	MCMC for Poisson Regression
MCMCpoissonChange()	MCMC for a Poisson Regression Changepoint Model
MCMCprobit()	MCMC for Probit Regression
MCMCquantreg()	Bayesian quantile regression using Gibbs sampling
MCMCregress()	MCMC for Gaussian Linear Regression
MCMCtobit()	MCMC for Gaussian Linear Regression with a Censored-Dependent Variable

Table 5.2: Bayesian Modeling Functions Available within the `MCMCpack` Library

5.5 Multivariate statistics and discriminant procedures

This section includes a sampling of commonly used multivariate, clustering methods, and discriminant procedures [42, 67]. The CRAN Task Views on multivariate statistics, cluster analysis and psychometrics provide additional descriptions of available functionality.

5.5.1 Cronbach's alpha

Example: See 5.7.17

Cronbach's α is a statistic that summarizes the internal consistency and reliability of a set of items comprising a measure.

```
library(multilevel)
cronbach(cbind(x1, x2, ..., xk))
```

5.5.2 Factor analysis

Example: See 5.7.18

Factor analysis is used to explain variability of a set of measures in terms of underlying unobservable factors. The observed measures can be expressed as linear combinations of the factors, plus random error. Factor analysis is often used as a way to guide the creation of summary scores from individual items.

```
res = factanal(~ x1 + x2 + ... + xk, factors=nfact)
print(res, cutoff=cutoffval, sort=TRUE)
```

By default no scores are produced (this is controlled with the `scores` option). A rotation function must be specified: options include `varimax()` and `promax()`. When printing `factanal` objects, values less than the specified cutoff (in absolute value) are not displayed.

5.5.3 Principal component analysis

Example: See 5.7.19

Principal component analysis is a data reduction technique which can create uncorrelated variables which account for some of the variability in a set of variables.

```
newds = na.omit(data.frame(x1, x2, ..., xk))
pcavals = prcomp(newds, scale=TRUE)
summary(pcavals)
```

The `biplot()` command can be used to graphically present the results, with the `choices` option allowing specification of which factors to plot.

5.5.4 Recursive partitioning

Example: See 5.7.20

Recursive partitioning is used to create a decision tree to classify observations from a dataset based on categorical predictors. This functionality is available within the `rpart` package.

```
library(rpart)
rpartout = rpart(x1 ~ x2 + x3 + ... + xk, method="class",
   data=ds)
printcp(rpartout)
plot(rpartout)
text(rpartout)
```

Supported methods include `anova`, `class`, `exp`, or `poisson`. The `printcp()`, `plot()`, and `text()` functions operate on `rpart` class objects.

5.5.5 Linear discriminant analysis

Example: See 5.7.21

Linear (or Fisher) discriminant analysis is used to find linear combinations of variables that can separate classes.

```
library(MASS)
ngroups = length(unique(y))
ldamodel = lda(y ~ x1 + ... + xk, prior=rep(1/ngroups, ngroups),
   data=ds)
print(ldamodel)
plot(ldamodel)
```

The prior probabilities of class membership can be left unspecified, or given in the order of the factor levels. Details on display of `lda` objects can be found using `help(plot.lda)`.

5.5.6 Hierarchical clustering

Example: See 5.7.22

Many techniques exist for grouping similar variables or similar observations. These groups, or clusters, can be overlapping or disjoint, and are sometimes placed in a hierarchical structure so that some disjoint clusters share a higher-level cluster. Clustering tools available include `hclust()` and `kmeans()`. The function `dendrogram()`, also in the `stats` package, plots tree diagrams. The `cluster()` package contains functions `pam()`, `clara()`, and `diana()`. The CRAN Clustering Task View has more details (see also 6.6.9, visualizing correlation matrices).

```
cormat = cor(cbind(x1, x2, ..., xk), use="pairwise.complete.obs")
hclustobj = hclust(dist(cormat))
```

5.6 Further resources

Many of the topics covered in this chapter are active areas of statistical research and many foundational articles are still useful. Here we provide references to texts which serve as accessible references.

Dobson and Barnett [9] presents an accessible introduction to generalized linear models, while McCullagh and Nelder's work [43] remains a classic. Agresti [1] describes the analysis of categorical data.

Fitzmaurice, Laird, and Ware [17] provide an accessible overview of mixed effects methods while West, Welch, and Galecki [78] review these methods for a variety of statistical packages. A comprehensive review of the material in this chapter is incorporated in Faraway's text [15]. The text by Hardin and Hilbe [23] provides a review of generalized estimating equations. The CRAN Task View on Analysis of Spatial Data provides a summary of tools to read, visualize, and analyze spatial data. Collett [5] presents an accessible introduction to survival analysis. Gelman, Carlin, Stern, Rubin [21] provide a comprehensive introduction to Bayesian inference, while Albert [2] focuses on use of R for Bayesian computations. Särndal, Swensson, and Wretman [62] provide a readable overview of the analysis of data from complex surveys, while Amico [8] describes the implementations within R.

Manly [42] and Tabachnick and Fidell [67] provide a comprehensive introduction to multivariate statistics. The CRAN Task View on Psychometric models and methods describe support for Rasch, item response, structural equation and related models, while the Multivariate Statistics View includes sections on visualization, testing, multivariate distributions, projection and scaling methods, classification, correspondence analysis, and latent variable approaches.

5.7 HELP examples

To help illustrate the tools presented in this chapter, we apply many of the entries to the HELP data. The code for these examples can be downloaded from `http://www.math.smith.edu/r/examples`.

```
> options(digits=3)
> options(width=67)
> options(show.signif.stars=FALSE)
> load("savedfile")  # this was created in chapter 2
> attach(ds)
```

We begin by loading (2.1.1) the HELP dataset using a saved file that was created in Section 2.13.3.

5.7.1 Logistic regression

In this example, we fit a logistic regression (5.1.1) to model the probability of being homeless (spending one or more nights in a shelter or on the street in the past six months) as a function of predictors. We use the `glm()` command to fit the logistic regression model.

```
> logres = glm(homeless ~ female + i1 + substance + sexrisk +
+    indtot, binomial, data=ds)
> names(logres)

 [1] "coefficients"     "residuals"      "fitted.values"
 [4] "effects"          "R"              "rank"
 [7] "qr"               "family"         "linear.predictors"
[10] "deviance"         "aic"            "null.deviance"
[13] "iter"             "weights"        "prior.weights"
[16] "df.residual"      "df.null"        "y"
[19] "converged"        "boundary"       "model"
[22] "call"             "formula"        "terms"
[25] "data"             "offset"         "control"
[28] "method"           "contrasts"      "xlevels"
> anova(logres)
Analysis of Deviance Table
Model: binomial, link: logit
Response: homeless
Terms added sequentially (first to last)


          Df Deviance Resid. Df Resid. Dev
NULL                     452       625
female     1    4.37    451       621
i1         1   25.79    450       595
substance  2    3.67    448       591
sexrisk    1    5.97    447       585
indtot     1    8.84    446       577
```

```
> summary(logres)

Call:
glm(formula = homeless ~ female + i1 + substance + sexrisk +
    indtot, family = binomial, data = ds)

Deviance Residuals:
   Min      1Q  Median      3Q     Max
 -1.75   -1.04   -0.70    1.13    2.03

Coefficients:
                  Estimate Std. Error z value Pr(>|z|)
(Intercept)       -2.13192    0.63347   -3.37  0.00076
female            -0.26170    0.25146   -1.04  0.29800
i1                 0.01749    0.00631    2.77  0.00556
substancecocaine  -0.50335    0.26453   -1.90  0.05707
substanceheroin   -0.44314    0.27030   -1.64  0.10113
sexrisk            0.07251    0.03878    1.87  0.06152
indtot             0.04669    0.01622    2.88  0.00399

(Dispersion parameter for binomial family taken to be 1)

    Null deviance: 625.28  on 452  degrees of freedom
Residual deviance: 576.65  on 446  degrees of freedom
AIC: 590.7

Number of Fisher Scoring iterations: 4
```

More information can be found in the summary() output object.

```
> names(summary(logres))

 [1] "call"           "terms"           "family"
 [4] "deviance"       "aic"             "contrasts"
 [7] "df.residual"    "null.deviance"   "df.null"
[10] "iter"           "deviance.resid"  "coefficients"
[13] "aliased"        "dispersion"      "df"
[16] "cov.unscaled"   "cov.scaled"
```

```
> summary(logres)$coefficients
```

```
                  Estimate Std. Error z value Pr(>|z|)
(Intercept)        -2.1319    0.63347   -3.37 0.000764
female             -0.2617    0.25146   -1.04 0.297998
i1                  0.0175    0.00631    2.77 0.005563
substancecocaine   -0.5033    0.26453   -1.90 0.057068
substanceheroin    -0.4431    0.27030   -1.64 0.101128
sexrisk             0.0725    0.03878    1.87 0.061518
indtot              0.0467    0.01622    2.88 0.003993
```

5.7.2   Poisson regression

In this example we fit a Poisson regression model (5.1.3) for i1, the average number of drinks per day in the 30 days prior to entering the detox center.

```
> poisres = glm(i1 ~ female + substance + age, poisson)
> summary(poisres)

Call:
glm(formula = i1 ~ female + substance + age, family = poisson)

Deviance Residuals:
   Min     1Q  Median     3Q    Max
 -7.57  -3.69   -1.40   1.04  15.99

Coefficients:
                  Estimate Std. Error z value Pr(>|z|)
(Intercept)        2.89785    0.05827   49.73  < 2e-16
female            -0.17605    0.02802   -6.28  3.3e-10
substancecocaine  -0.81715    0.02776  -29.43  < 2e-16
substanceheroin   -1.12117    0.03392  -33.06  < 2e-16
age                0.01321    0.00145    9.08  < 2e-16

(Dispersion parameter for poisson family taken to be 1)

    Null deviance: 8898.9  on 452   degrees of freedom
Residual deviance: 6713.9  on 448   degrees of freedom
AIC: 8425

Number of Fisher Scoring iterations: 6
```

It is always important to check assumptions for models. This is particularly true for Poisson models, which are quite sensitive to model departures. There is support for Pearson's χ^2 goodness of fit test.

```
> library(vcd)
> poisfit = goodfit(e2b, "poisson")
> summary(poisfit)

          Goodness-of-fit test for poisson distribution

                  X^2 df P(> X^2)
Likelihood Ratio 208 10   3.6e-39
```

The results indicate that the fit is poor ($\chi^2_{10} = 208$, $p < 0.0001$); the Poisson model does not appear to be tenable.

5.7.3 Zero-inflated Poisson regression

A zero-inflated Poisson regression model (5.1.5) might fit better.

```
> library(pscl)

Classes and Methods for R developed in the
Political Science Computational Laboratory
Department of Political Science
Stanford University
Simon Jackman
hurdle and zeroinfl functions by Achim Zeileis

> res = zeroinfl(i1 ~ female + substance + age | female, data=ds)
> res

Call:
zeroinfl(formula = i1 ~ female + substance + age | female,
    data = ds)

Count model coefficients (poisson with log link):
      (Intercept)            female  substancecocaine
          3.05781          -0.06797          -0.72466
   substanceheroin               age
         -0.76086           0.00927

Zero-inflation model coefficients (binomial with logit link):
(Intercept)         female
     -1.979          0.843
```

Women are more likely to abstain from alcohol than men (p=0.0025), as well as drink less when they drink (p=0.015). Other significant predictors include substance and age, though model assumptions for count models should always be carefully verified [26].

5.7.4   Negative binomial regression

A negative binomial regression model (5.1.6) might improve on the Poisson.

```
> library(MASS)
> nbres = glm.nb(i1 ~ female + substance + age)
> summary(nbres)

Call:
glm.nb(formula = i1 ~ female + substance + age,
    init.theta = 0.810015139, link = log)

Deviance Residuals:
   Min      1Q  Median      3Q     Max
-2.414  -1.032  -0.278   0.241   2.808

Coefficients:
                   Estimate Std. Error z value Pr(>|z|)
(Intercept)         3.01693    0.28928   10.43  < 2e-16
female             -0.26887    0.12758   -2.11    0.035
substancecocaine   -0.82360    0.12904   -6.38  1.7e-10
substanceheroin    -1.14879    0.13882   -8.28  < 2e-16
age                 0.01072    0.00725    1.48    0.139

(Dispersion parameter for Negative Binomial(0.81) family
taken to be 1)
    Null deviance: 637.82  on 452  degrees of freedom
Residual deviance: 539.60  on 448  degrees of freedom
AIC: 3428

Number of Fisher Scoring iterations: 1


            Theta:  0.8100
        Std. Err.:  0.0589

 2 x log-likelihood:  -3416.3340
```

The deviance / DF is close to 1, suggesting a reasonable fit.

5.7.5   LASSO model selection

In this section, we can provide guidance on selecting a regression model for the
CESD score as a function of a large number of predictors (e.g., in a data mining
setting) using the lasso method (5.4.4). Here we illustrate the technique with
a set of 5 predictors.

```
> library(lars)
> lassores = lars(cbind(female, i1, age, mcs, pcs, homeless),
+     cesd, type="lasso")
> print(lassores)

Call:
lars(x=cbind(female, i1, age, mcs, pcs, homeless), y=cesd,
    type = "lasso")
R-squared: 0.526
Sequence of LASSO moves:
     mcs pcs i1 female age homeless
Var    4   5  2      1   3        6
Step   1   2  3      4   5        6

> coef(lassores)

       female      i1     age     mcs     pcs homeless
[1,]     0.00 0.00000  0.0000   0.000   0.000    0.000
[2,]     0.00 0.00000  0.0000  -0.426   0.000    0.000
[3,]     0.00 0.00000  0.0000  -0.552  -0.149    0.000
[4,]     0.00 0.00522  0.0000  -0.561  -0.159    0.000
[5,]     1.81 0.04367  0.0000  -0.611  -0.207    0.000
[6,]     1.98 0.04802 -0.0143  -0.615  -0.213    0.000
[7,]     2.45 0.05743 -0.0518  -0.624  -0.229    0.342

> summary(lassores)

LARS/LASSO
Call: lars(x=cbind(female, i1, age, mcs, pcs, homeless), y=cesd,
Call:      type = "lasso")
  Df  Rss      Cp
0  1 70788  490.10
1  2 42098  110.67
2  3 35796   28.89
3  4 35448   26.27
4  5 33727    5.38
5  6 33647    6.31
6  7 33548    7.00
```

The estimated standardized coefficients are displayed in Figure 5.1. If the constraint was set to 0.5, only variables 4 and 5 (MCS and PCS) would be included in the model.

```
> plot(lassores)
```

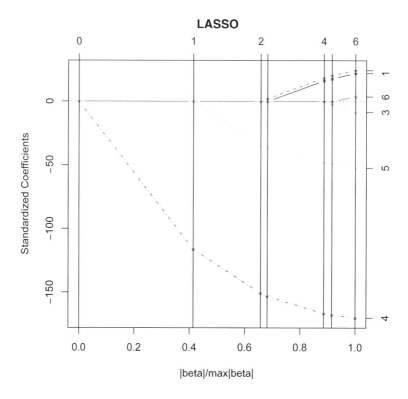

Figure 5.1: LASSO regression results.

5.7.6   Quantile regression

In this section, we fit a quantile regression model (5.4.5) of the number of drinks (i1) as a function of predictors, modeling the 75th percentile (Q3).

```
> library(quantreg)
> quantres = rq(i1 ~ female + substance + age, tau=0.75,
+     data=ds)
```

```
> summary(quantres)

Call: rq(formula = i1 ~ female + substance + age, tau=0.75)

tau: [1] 0.75

Coefficients:
                    coefficients lower bd upper bd
(Intercept)            29.636      14.150   42.603
female                 -2.909      -7.116    3.419
substancecocaine      -20.091     -29.011  -15.460
substanceheroin       -22.636     -28.256  -19.115
age                     0.182      -0.153    0.468

> detach("package:quantreg")
```

Because the quantreg package overrides needed functionality in other packages, we detach() it after running the rq() function (see 1.5.6).

5.7.7   Ordinal logit

To demonstrate an ordinal logit analysis (5.1.9), we first create an ordinal categorical variable from the sexrisk variable, then model this three-level ordinal variable as a function of cesd and pcs.

```
> library(MASS)
> sexriskcat = as.factor(as.numeric(sexrisk>=2) +
+    as.numeric(sexrisk>=6))
> table(sexriskcat)

sexriskcat
  0   1   2
 58 244 151
```

```
> ologit = polr(sexriskcat ~ cesd + pcs)
> summary(ologit)

Call:
polr(formula = sexriskcat ~ cesd + pcs)

Coefficients:
         Value Std. Error  t value
cesd -3.72e-05    0.00762 -0.00489
pcs   5.23e-03    0.00876  0.59648

Intercepts:
    Value  Std. Error t value
0|1 -1.669  0.562       -2.971
1|2  0.944  0.556        1.698

Residual Deviance: 871.76
AIC: 879.76
```

5.7.8   Multinomial logit

We can fit a multinomial logit (5.1.10) model for the categorized `sexrisk` variable.

```
> library(VGAM)
> mlogit = vglm(sexriskcat ~ cesd + pcs, family=multinomial())
```

```
> summary(mlogit)

Call:
vglm(formula = sexriskcat ~ cesd + pcs, family = multinomial())

Pearson Residuals:
                        Min   1Q Median   3Q Max
log(mu[,1]/mu[,3]) -0.8 -0.7   -0.2 -0.1   3
log(mu[,2]/mu[,3]) -1.3 -1.2    0.8  0.9   1

Coefficients:
               Value Std. Error t value
(Intercept):1 -0.686      0.948    -0.7
(Intercept):2  0.791      0.639     1.2
cesd:1         0.007      0.013     0.5
cesd:2        -0.007      0.009    -0.8
pcs:1         -0.010      0.015    -0.7
pcs:2         -0.002      0.010    -0.2

Number of linear predictors:  2

Names of linear predictors:
log(mu[,1]/mu[,3]), log(mu[,2]/mu[,3])

Dispersion Parameter for multinomial family:   1

Residual Deviance: 870 on 900 degrees of freedom

Log-likelihood: -435 on 900 degrees of freedom

Number of Iterations: 4

> detach("package:VGAM")
```

Because the VGAM package overrides needed functionality in other packages, we detach() it after running the vglm() function (see 1.5.6).

5.7.9   Generalized additive model

We can fit a generalized additive model (5.4.2), and generate a plot.

```
> library(gam)
> gamreg= gam(cesd ~ female + lo(pcs) + substance)
```

```
> summary(gamreg)

Call: gam(formula = cesd ~ female + lo(pcs) + substance)
Deviance Residuals:
   Min    1Q Median    3Q    Max
-29.16  -8.14   0.81   8.23  29.25


(Dispersion Parameter for gaussian family taken to be 135)


    Null Deviance: 70788 on 452 degrees of freedom
Residual Deviance: 60288 on 445 degrees of freedom
AIC: 3519

Number of Local Scoring Iterations: 2

DF for Terms and F-values for Nonparametric Effects

            Df Npar Df Npar F Pr(F)
(Intercept)  1
female       1
lo(pcs)      1    3.1    3.77 0.010
substance    2

> coefficients(gamreg)

    (Intercept)            female           lo(pcs)
         46.524             4.339            -0.277
substancecocaine   substanceheroin
         -3.956            -0.205
```

The estimated smoothing function is displayed in Figure 5.2.

```
> plot(gamreg, terms=c("lo(pcs)"), se=2, lwd=3)
> abline(h=0)
```

5.7.10 Reshaping dataset for longitudinal regression

A wide (multivariate) dataset can be reshaped (2.5.3) into a tall (longitudinal)
dataset. Here we create time-varying variables (with a suffix tv) as well as keep
baseline values (without the suffix). There are four lines in the long dataset for
every line in the original dataset. We use the reshape() command to transpose
the dataset.

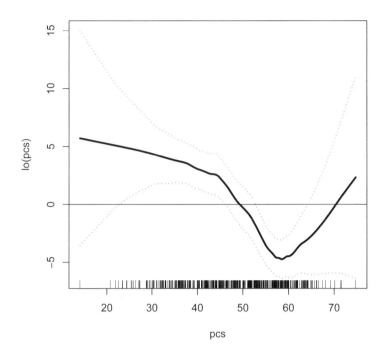

Figure 5.2: Plot of the smoothed association between PCS and CESD.

```
> long = reshape(ds, idvar="id",
+     varying=list(c("cesd1","cesd2","cesd3","cesd4"),
+     c("mcs1","mcs2","mcs3","mcs4"), c("i11","i12","i13","i14"),
+     c("g1b1","g1b2","g1b3","g1b4")),
+     v.names=c("cesdtv","mcstv","i1tv","g1btv"),
+     timevar="time", times=1:4, direction="long")
> detach(ds)
```

We can check the resulting dataset by printing tables by time,

```
> table(long$g1btv, long$time)

      1   2   3   4
  0 219 187 225 245
  1  27  22  22  21
```

or by looking at the observations over time for a given individual.

```
> attach(long)
> long[id==1, c("id", "time", "cesd", "cesdtv")]

    id time cesd cesdtv
1.1  1    1   49      7
1.2  1    2   49     NA
1.3  1    3   49      8
1.4  1    4   49      5

> detach(long)
```

This process can be reversed, creating a wide dataset from a tall one with another call to reshape().

```
> wide = reshape(long,
+     v.names=c("cesdtv", "mcstv", "i1tv", "g1btv"),
+     idvar="id", timevar="time", direction="wide")
> wide[c(2,8), c("id", "cesd", "cesdtv.1", "cesdtv.2",
+     "cesdtv.3", "cesdtv.4")]

    id cesd cesdtv.1 cesdtv.2 cesdtv.3 cesdtv.4
2.1  2   30       11       NA       NA       NA
8.1  8   32       18       NA       25       NA
```

5.7.11   Linear model for correlated data

Here we fit a general linear model for correlated data (modeling the covariance matrix directly, 5.2.1). In this example, the estimated correlation matrix for the seventh subject is printed (this subject was selected because all four time points were observed).

```
> library(nlme)
> glsres = gls(cesdtv ~ treat + as.factor(time),
+     correlation=corSymm(form = ~ time | id),
+     weights=varIdent(form = ~ 1 | time), long,
+     na.action=na.omit)
```

```
> summary(glsres)

Generalized least squares fit by REML
  Model: cesdtv ~ treat + as.factor(time)
  Data: long
   AIC  BIC logLik
  7550 7623  -3760


Correlation Structure: General
 Formula: ~time | id
 Parameter estimate(s):
 Correlation:
  1     2     3
2 0.584
3 0.639 0.743
4 0.474 0.585 0.735
Variance function:
 Structure: Different standard deviations per stratum
 Formula: ~1 | time
 Parameter estimates:
    1     3     4     2
1.000 0.996 0.996 1.033


Coefficients:
                 Value Std.Error t-value p-value
(Intercept)      23.66    1.098   21.55   0.000
treat            -0.48    1.320   -0.36   0.716
as.factor(time)2  0.28    0.941    0.30   0.763
as.factor(time)3 -0.66    0.841   -0.78   0.433
as.factor(time)4 -2.41    0.959   -2.52   0.012


 Correlation:
                 (Intr) treat  as.()2 as.()3
treat            -0.627
as.factor(time)2 -0.395  0.016
as.factor(time)3 -0.433  0.014  0.630
as.factor(time)4 -0.464  0.002  0.536  0.708


Standardized residuals:
   Min    Q1    Med    Q3    Max
-1.643 -0.874 -0.115  0.708  2.582


Residual standard error: 14.4
Degrees of freedom: 969 total; 964 residual
```

```
> anova(glsres)

Denom. DF: 964
              numDF F-value p-value
(Intercept)       1    1168  <.0001
treat             1       0  0.6887
as.factor(time)   3       4  0.0145
```

A set of parallel boxplots (6.1.12) by time can be generated using the following
commands. Results are displayed in Figure 5.3.

```
> library(lattice)
> bwplot(cesdtv ~ as.factor(treat)| time, xlab="TREAT",
+    strip=strip.custom(strip.names=TRUE, strip.levels=TRUE),
+    ylab="CESD", layout=c(4,1), col="black", data=long,
+    par.settings=list(box.rectangle=list(col="black"),
+       box.dot=list(col="black"),
+       box.umbrella=list(col="black")))
```

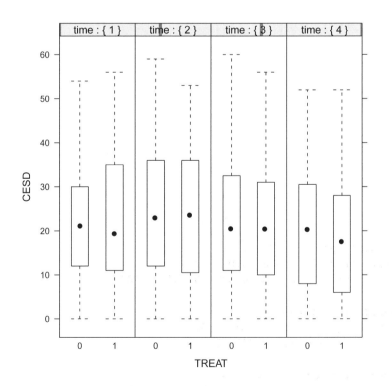

Figure 5.3: Side-by-side boxplots of CESD by treatment and time.

5.7.12   Linear mixed (random slope) model

Here we fix a mixed effects, or random slope model (5.2.3). We specify a categorical fixed effect of time but a random slope across time treated continuously. We begin by creating an `as.factor()` version of time. As an alternative, we could nest the call to `as.factor()` within the call to `lme()`.

```
> attach(long)
> tf = as.factor(time)
> library(nlme)
> lmeslope = lme(fixed=cesdtv ~ treat + tf,
+   random=~ time |id, data=long, na.action=na.omit)
> print(lmeslope)

Linear mixed-effects model fit by REML
  Data: long
  Log-restricted-likelihood: -3772
  Fixed: cesdtv ~ treat + tf
(Intercept)      treat        tf2        tf3        tf4
    23.8843    -0.4353    -0.0614    -1.0142    -2.5776

Random effects:
 Formula: ~time | id
 Structure: General positive-definite, Log-Cholesky param
             StdDev Corr
(Intercept) 13.73  (Intr)
time         3.03  -0.527
Residual     7.85

Number of Observations: 969
Number of Groups: 383
```

```
> anova(lmeslope)

              numDF denDF F-value p-value
(Intercept)       1   583    1163  <.0001
treat             1   381       0  0.7257
tf                3   583       3  0.0189
```

The `random.effects()` and `predict()` functions are used to find the predicted random effects and predicted values, respectively.

```
> reffs = random.effects(lmeslope)
> reffs[1,]

  (Intercept)    time
1       -13.5 -0.0239
```

```
> predval = predict(lmeslope, newdata=long, level=0:1)
> predval[id==1,]

    id predict.fixed predict.id
1.1  1          23.4       9.94
1.2  1          23.4       9.86
1.3  1          22.4       8.88
1.4  1          20.9       7.30

> vc = VarCorr(lmeslope)
> summary(vc)

   Variance    StdDev       Corr
    9.17:1      3.03:1        :1
   61.58:1      7.85:1   -0.527:1
  188.43:1     13.73:1   (Intr):1

> detach(long)
```

The VarCorr() function calculates the variances, standard deviations and correlations between the random effects terms, as well as the within-group error variance and standard deviation.

5.7.13   Generalized estimating equations

We fit a generalized estimating equation (GEE) model (5.2.7), using an exchangeable working correlation matrix and empirical variance [39]. Note that in the current release of the gee package, unstructured working correlations are not supported with nonmonotone missingness.

```
> library(gee)
> sortlong = long[order(long$id),]
> attach(sortlong)
> form = formula(g1btv ~ treat + time)
> geeres = gee(formula = form, id=id, data=sortlong,
+    family=binomial, na.action=na.omit,
+    corstr="exchangeable")

(Intercept)       treat        time
    -1.9649      0.0443     -0.1256
```

In addition to returning an object with results, the gee() function displays the coefficients from a model assuming that all observations are uncorrelated.

```
> coef(geeres)

(Intercept)         treat          time
   -1.85169      -0.00874      -0.14593

> sqrt(diag(geeres$robust.variance))

(Intercept)         treat          time
    0.2723        0.2683        0.0872

> geeres$working.correlation

       [,1]   [,2]   [,3]   [,4]
[1,]  1.000  0.299  0.299  0.299
[2,]  0.299  1.000  0.299  0.299
[3,]  0.299  0.299  1.000  0.299
[4,]  0.299  0.299  0.299  1.000
```

5.7.14 Generalized linear mixed model

Here we fit a generalized linear mixed model (GLMM) (5.2.6), predicting recent suicidal ideation as a function of treatment, depressive symptoms (CESD) and time. Each subject is assumed to have their own random intercept.

```
> library(lme4)
> glmmres = lmer(g1btv ~ treat + cesdtv + time + (1|id),
+    family=binomial(link="logit"), data=long)
> summary(glmmres)
```

```
Generalized linear mixed model fit by the Laplace approximation
Formula: g1btv ~ treat + cesdtv + time + (1 | id)
   Data: long
 AIC BIC logLik deviance
 480 504   -235     470
Random effects:
 Groups Name        Variance Std.Dev.
 id     (Intercept) 32.6     5.71
Number of obs: 968, groups: id, 383

Fixed effects:
            Estimate Std. Error z value Pr(>|z|)
(Intercept)  -8.7632     1.2802   -6.85  7.6e-12
treat        -0.0417     1.2159   -0.03     0.97
cesdtv        0.1018     0.0237    4.30  1.7e-05
time         -0.2426     0.1837   -1.32     0.19

Correlation of Fixed Effects:
       (Intr) treat  cesdtv
treat  -0.480
cesdtv -0.641 -0.025
time   -0.366  0.009  0.028
```

5.7.15　Cox proportional hazards model

We fit a proportional hazards model (5.3.1) for the time to linkage to primary care, with randomization group, age, gender, and CESD as predictors. Here we request the Breslow estimator (the default is the Efron estimator).

```
> library(survival)
> survobj = coxph(Surv(dayslink, linkstatus) ~ treat + age +
+     female + cesd, method="breslow", data=ds)
```

```
> print(survobj)

Call:
coxph(formula = Surv(dayslink, linkstatus) ~ treat + age +
    female + cesd, data = ds, method = "breslow")


          coef exp(coef) se(coef)      z     p
treat   1.65186     5.217  0.19324  8.548 0.000
age     0.02467     1.025  0.01032  2.391 0.017
female -0.32535     0.722  0.20379 -1.597 0.110
cesd    0.00235     1.002  0.00638  0.369 0.710

Likelihood ratio test=94.6  on 4 df, p=0  n=431 (22 observations
deleted due to missingness)
```

5.7.16   Bayesian Poisson regression

In this example, we fit a Poisson regression model to the count of alcohol drinks in the HELP study as fit previously (5.7.2), this time using Markov Chain Monte Carlo methods (5.4.7).

```
> ds = read.csv("http://www.math.smith.edu/r/data/help.csv")
> attach(ds)
> library(MCMCpack)
Loading required package: coda
Loading required package: lattice
Loading required package: MASS
##
## Markov Chain Monte Carlo Package (MCMCpack)
## Copyright (C) 2003-2008 Andrew D. Martin, Kevin M. Quinn,
## and Jonh Hee Park
## Support provided by the U.S. National Science Foundation
## (Grants SES-0350646 and SES-0350613)
##
> posterior = MCMCpoisson(i1 ~ female + as.factor(substance) +
    age)
The Metropolis acceptance rate for beta was 0.27891
```

```
> summary(posterior)
Iterations = 1001:11000
Thinning interval = 1
Number of chains = 1
Sample size per chain = 10000
1. Empirical mean and standard deviation for each variable,
   plus standard error of the mean:
                          Mean       SD Naive SE     TS-SE
(Intercept)             2.8959 0.05963 5.96e-04  0.002858
female                 -0.1752 0.02778 2.78e-04  0.001085
as.factor(subs)cocaine -0.8176 0.02727 2.73e-04  0.001207
as.factor(subs)heroin  -1.1199 0.03430 3.43e-04  0.001333
age                     0.0133 0.00148 1.48e-05  0.000071

2. Quantiles for each variable:
                          2.5%     25%     50%     75%   97.5%
(Intercept)             2.7807  2.8546  2.8952  2.9390  3.0157
female                 -0.2271 -0.1944 -0.1754 -0.1567 -0.1184
as.factor(subs)cocaine -0.8704 -0.8364 -0.8174 -0.7992 -0.7627
as.factor(subs)heroin  -1.1858 -1.1430 -1.1193 -1.0967 -1.0505
age                     0.0103  0.0122  0.0133  0.0143  0.0160
```

Default plots are available for MCMC objects returned by MCMCpack. These can be displayed using the command plot(posterior).

5.7.17   Cronbach's alpha

We begin by calculating Cronbach's α for the 20 items comprising the CESD (Center for Epidemiologic Studies–Depression scale).

```
> library(multilevel)
> cronbach(cbind(f1a, f1b, f1c, f1d, f1e, f1f, f1g, f1h, f1i,
+     f1j, f1k, f1l, f1m, f1n, f1o, f1p, f1q, f1r, f1s, f1t))

$Alpha
[1] 0.761


$N
[1] 446
```

The observed α of 0.76 from the HELP study is relatively low: this may be due to ceiling effects for this sample of subjects recruited in a detoxification unit.

5.7.18   Factor analysis

Here we consider a maximum likelihood factor analysis with varimax rotation for the individual items of the `cesd` (Center for Epidemiologic Studies–Depression) scale. The individual questions can be found in Table A.2 in the Appendix. We arbitrarily force three factors.

```
> res = factanal(~ f1a + f1b + f1c + f1d + f1e + f1f + f1g +
+    f1h + f1i + f1j + f1k + f1l + f1m + f1n + f1o + f1p + f1q +
+    f1r + f1s + f1t, factors=3, rotation="varimax",
+    na.action=na.omit, scores="regression")
```

```
> print(res, cutoff=0.45, sort=TRUE)
Call:
factanal(x = ~f1a + f1b + f1c + f1d + f1e + f1f + f1g + f1h +
    f1i + f1j + f1k + f1l + f1m + f1n + f1o + f1p + f1q + f1r +
    f1s + f1t, factors = 3, na.action = na.omit,
    scores = "regression", rotation = "varimax")
Uniquenesses:
  f1a   f1b   f1c   f1d   f1e   f1f   f1g   f1h   f1i   f1j
0.745 0.768 0.484 0.707 0.701 0.421 0.765 0.601 0.616 0.625
  f1k   f1l   f1m   f1n   f1o   f1p   f1q   f1r   f1s   f1t
0.705 0.514 0.882 0.623 0.644 0.407 0.713 0.467 0.273 0.527
Loadings:
    Factor1 Factor2 Factor3
f1c  0.618
f1e  0.518
f1f  0.666
f1k  0.523
f1r  0.614
f1h         -0.621
f1l         -0.640
f1p         -0.755
f1o                  0.532
f1s                  0.802
f1a
f1b
f1d         -0.454
f1g  0.471
f1i  0.463
f1j  0.495
f1m
f1n  0.485
f1q  0.457
f1t  0.489
```

```
                Factor1 Factor2 Factor3
SS loadings       3.847   2.329   1.636
Proportion Var    0.192   0.116   0.082
Cumulative Var    0.192   0.309   0.391


Test of the hypothesis that 3 factors are sufficient.
The chi square statistic is 289 on 133 degrees of freedom.
The p-value is 1.56e-13
```

The item loadings match our intuitions. We see that the second factor loads on the reverse-coded items (H, L, P, and D, see 2.13.5). Factor 3 loads on items O and S (*people were unfriendly* and *I felt that people dislike me*).

5.7.19   Principal component analysis

Here we estimate the principal components for the variables representing the mental component score, CESD score, drug risk, sex risk, and inventory of drug use consequences.

```
> ds = read.csv("http://www.math.smith.edu/r/data/help.csv")
> attach(ds)
> smallds = na.omit(data.frame(-mcs, cesd1, drugrisk, sexrisk,
  indtot))
> pcavals = prcomp(smallds, scale=TRUE)

> pcavals
Standard deviations:
[1] 1.308 1.008 0.979 0.883 0.731


Rotation:
            PC1      PC2     PC3     PC4     PC5
X.mcs    -0.621   0.0522   0.135   0.104   0.7634
cesd1    -0.416   0.3595   0.684  -0.075  -0.4734
drugrisk -0.394   0.3055  -0.576  -0.629  -0.1541
sexrisk  -0.222  -0.8482   0.196  -0.428  -0.0984
indtot   -0.487  -0.2351  -0.379   0.636  -0.3996
> summary(pcavals)
Importance of components:
                        PC1    PC2    PC3    PC4    PC5
Standard deviation     1.308  1.008  0.979  0.883  0.731
Proportion of Variance 0.342  0.203  0.192  0.156  0.107
Cumulative Proportion  0.342  0.546  0.737  0.893  1.000
```

The `biplot()` command can be used to graphically present the results. Figure 5.4 displays these values for the first and second component, with a small size specified for the observations.

```
> biplot(pcavals, choices=c(1,2), cex=c(0.4, 1))
```

The first component appears to be associated with all of the variables except
sexrisk, while the second component is associated primarily with that variable.

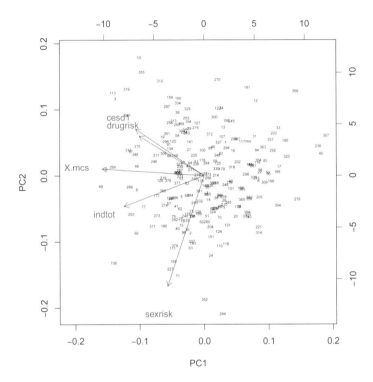

Figure 5.4: Biplot of first and second principal component.

5.7.20   Recursive partitioning

In this example, we attempt to classify subjects based on their homeless status,
using gender, drinking, primary substance, RAB sex risk, MCS, and PCS as
predictors.

```
> library(rpart)
> sub = as.factor(substance)
> homeless.rpart = rpart(homeless ~ female + i1 + sub +
+      sexrisk + mcs + pcs, method="class", data=ds)
```

```
> printcp(homeless.rpart)

Classification tree:
rpart(formula = homeless ~ female + i1 + sub + sexrisk + mcs +
    pcs, data = ds, method = "class")

Variables actually used in tree construction:
[1] female  i1        mcs       pcs       sexrisk

Root node error: 209/453 = 0.5

n= 453

      CP nsplit rel error xerror xstd
1 0.10      0       1.0      1 0.05
2 0.05      1       0.9      1 0.05
3 0.03      4       0.8      1 0.05
4 0.02      5       0.7      1 0.05
5 0.01      7       0.7      1 0.05
6 0.01      9       0.7      1 0.05
```

Figure 5.5 displays the tree.

```
> plot(homeless.rpart)
> text(homeless.rpart)
```

To help interpret this model, we can assess the proportion of homeless among those with i1< 3.5 by pcs divided at 31.94.

```
> home = homeless[i1<3.5]
> pcslow = pcs[i1<3.5]<=31.94
> table(home, pcslow)

     pcslow
home FALSE TRUE
   0    89    2
   1    31    5

> rm(home, pcslow)
```

Among this subset, 71.4% (5 of 7) of those with low PCS scores are homeless, while only 25.8% (31 of 120) of those with PCS scores above the threshold are homeless.

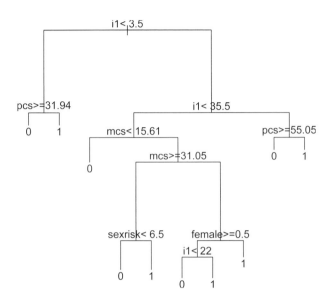

Figure 5.5: Recursive partitioning tree.

5.7.21 Linear discriminant analysis

We use linear discriminant analysis to distinguish between homeless and non-homeless subjects, with a prior expectation that half are in each group.

```
> library(MASS)
> ngroups = length(unique(homeless))
> ldamodel = lda(homeless ~ age + cesd + mcs + pcs,
+     prior=rep(1/ngroups, ngroups))
```

```
> print(ldamodel)

Call:
lda(homeless ~ age + cesd + mcs + pcs, prior = rep(1/ngroups,
    ngroups))

Prior probabilities of groups:
  0   1
0.5 0.5

Group means:
    age cesd  mcs  pcs
0 35.0 31.8 32.5 49.0
1 36.4 34.0 30.7 46.9

Coefficients of linear discriminants:
          LD1
age    0.0702
cesd   0.0269
mcs   -0.0195
pcs   -0.0426
```

The results indicate that homeless subjects tend to be older, have higher CESD scores, and lower MCS and PCS scores. Figure 5.6 displays the distribution of linear discriminant function values by homeless status. The distribution of the linear discriminant function values are shifted to the right for the homeless subjects, though there is considerable overlap between the groups.

```
> plot(ldamodel)
```

5.7.22   Hierarchical clustering

In this example, we cluster continuous variables from the HELP dataset.

```
> cormat = cor(cbind(mcs, pcs, cesd, i1, sexrisk),
+    use="pairwise.complete.obs")
> hclustobj = hclust(dist(cormat))
```

Figure 5.7 displays the clustering. Not surprisingly, the MCS and PCS variables cluster together, since they both utilize similar questions and structures. The cesd and i1 variables cluster together, while there is a separate node for sexrisk.

```
> plot(hclustobj)
```

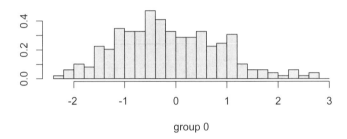

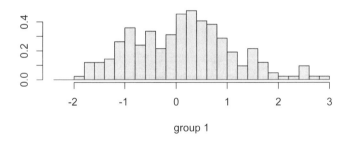

Figure 5.6: Graphical display of assignment probabilities or score functions from linear discriminant analysis by actual homeless status.

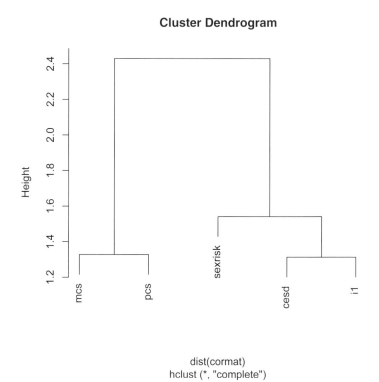

Figure 5.7: Results from hierarchical clustering.

Chapter 6

Graphics

This chapter describes how to create graphical displays, such as scatterplots, boxplots, and histograms. We provide a broad overview of the key ideas and techniques that are available. Additional discussion of ways to annotate displays and change defaults to present publication quality figures is included, as are details regarding how to output graphics in a variety of file formats (Section 6.4). Because graphics are useful to visualize analyses, examples appear throughout the HELP sections at the end of most of the chapters of the book.

Producing graphics for data analysis is simple and direct. Creating a graph for publication is more complex and typically requires a great deal of time to achieve the desired appearance. Our intent is to provide sufficient guidance that most effects can be achieved, but further investigation of the documentation and experimentation will doubtless be necessary for specific needs. There are a huge number of options: we aim to provide a road map as well as examples to illustrate the power of the package.

While many graphics can be generated using one command, figures are often built up element by element. For example, an empty box can be created with a specific set of x and y axis labels and tick marks, then points can be added with different printing characters. Text annotations can then be added, along with legends and other additional information. The Graphics Task View (http://cran.r-project.org/web/views) provides a comprehensive listing of functionality to create graphics.

A somewhat intimidating set of options is available, some of which can be specified using the `par()` graphics parameters (see Section 6.3), while others can be given as options to plotting commands (such as `plot()` or `lines()`).

A number of graphics devices support different platforms and formats. The default varies by platform (`Windows()` under Windows, `X11()` under Linux and `quartz()` under modern Mac OS X distributions). A device is created automatically when a plotting command is run, or a device can be started in advance to create a file in a particular format (e.g., the `pdf()` device).

A series of powerful add-on packages to create sophisticated graphics are available. These include the `grid` package [45], the `lattice` library [61], the `ggplot2` library and the `ROCR` package for receiver operating characteristic curves [66]. Running `example()` for a specified function of interest is particularly helpful for commands shown in this chapter, as is `demo(graphics)`.

6.1   A compendium of useful plots

6.1.1   Scatterplot

Example: See 4.7.1

See also 6.1.2 (scatterplot with multiple y values) and 6.1.5 (matrix of scatterplots)

```
plot(x, y)
```

Many objects have default plotting functions (e.g., for a linear model object, `plot.lm()` is called). More information can be found using `methods(plot)`. Specifying `type="n"` causes nothing to be plotted (but sets up axes and draws boxes, see 2.13.7). This technique is often useful if a plot is built up part by part. The `matplot()`, `matpoints()`, and `matlines()` functions can be used to plot columns of matrices against each other.

6.1.2   Scatterplot with multiple y values

See also 6.1.5 (matrix of scatterplots)   *Example:* See 6.6.1

```
plot(x, y1, pch=pchval1)    # create 1 plot with single y-axis
points(x, y2, pch=pchval2)
...
points(x, yk, pch=pchvalk)
```

or

```
# create 1 plot with 2 separate y axes
addsecondy = function(x, y, origy, yname="Y2") {
   prevlimits = range(origy)
   axislimits = range(y)
   axis(side=4, at=prevlimits[1] + diff(prevlimits)*c(0:5)/5,
       labels=round(axislimits[1] + diff(axislimits)*c(0:5)/5, 1))
   mtext(yname, side=4)
   newy = (y-axislimits[1])/
           (diff(axislimits)/diff(prevlimits)) + prevlimits[1]
   points(x, newy, pch=2)
}
```

```
plottwoy = function(x, y1, y2, xname="X", y1name="Y1",
    y2name="Y2")
{
    plot(x, y1, ylab=y1name, xlab=xname)
    addsecondy(x, y2, y1, yname=y2name)
}
plottwoy(x, y1, y2, y1name="Y1", y2name="Y2")
```

To create a figure with a single y axis value, it is straightforward to repeatedly call `points()` or other functions to add elements.

In the second example, two functions `addsecondy()` and `plottwoy()` are defined to add points on a new scale and an appropriate axis on the right. This involves rescaling and labeling the second axis (`side=4`) with six tick marks, as well as rescaling the y2 variable.

6.1.3 Bubble plot

Example: See 6.6.2

A bubble plot is a trivariate plot in which the size of the circle plotted on the scatterplot axes is proportional to a third variable. (see also 6.2.14, plotting symbols).

```
symbols(x, y, circles=z)
```

The vector given by the `circles` option denotes the radii of the circles.

6.1.4 Sunflower plot

Sunflower plots [10] are designed to display multiple observations (overplotting) at the same plotting position by adding additional components to the plotting symbol based on how many are at that position. Another approach to this problem involves jittering data (see 6.2.3).

```
sunflowerplot(x, y)
```

6.1.5 Matrix of scatterplots

Example: See 6.6.8

```
pairs(data.frame(x1, ..., xk))
```

The `pairs()` function is quite flexible, since it calls user-specified functions to determine what to display on the lower triangle, diagonal, and upper triangle (see `examples(pairs)` for illustration of its capabilities).

6.1.6   Conditioning plot

A conditioning plot is used to display a scatterplot for each level of one or two classification variables, as below.

Example: See 6.6.3

```
library(lattice)
coplot(y ~ x1 | x2*x3)
```

The `coplot()` function displays plots of `y` and `x1`, stratified by `x2` and `x3`. All variables may be either numeric or factors.

6.1.7   Barplot

While not typically an efficient graphical display, there are times when a barplot is appropriate to display counts by groups.

```
barplot(table(x1, x2), legend=c("grp1", "grp2"), xlab="X2")
```

or

```
library(lattice)
barchart(table(x1, x2, x3))
```

The input for the `barplot()` function is given as the output of a one- or two-dimensional contingency table, while the `barchart()` function available in the `library(lattice)` supports three-dimensional tables (see `example(barplot)` and `example(barchart)`). A similar `dotchart()` function produces a horizontal slot for each group with a dot reflecting the frequency.

6.1.8   Dotplot

Example: See 6.6.5

```
dotplot(~ x1 + x2, pch=c("1","2"))
```

The `dotplot()` function in `library(lattice)` is useful for displaying labeled quantitative values [32]. In this example, two values are plotted for each level of the variable `y`.

6.1.9   Histogram

Example: See 3.6.1

The example in Section 3.6.1 demonstrates how to annotate a histogram with an overlaid normal or kernel density estimates. Similar estimates are available for all other supported densities (see Table 2.1).

```
hist(x)
```

The default behavior for a histogram is to display frequencies on the vertical axis; probability densities can be displayed using the `freq=FALSE` option. The default title is given by `paste("Histogram of" , x)` where x is the name of the variable being plotted; this can be changed with the `main` option.

6.1.10   Stem-and-leaf plot

Example: See 4.7.4

Stem-and-leaf plots are text-based graphics that are particularly useful to describe the distribution of small datasets.

```
stem(x)
```

The `scale` option can be used to increase or decrease the number of stems (default value is 1).

6.1.11   Boxplot

See also 6.1.12 (side-by-side boxplots)   *Example:* See 4.7.6 and 5.7.11

```
boxplot(x)
```

The `boxplot()` function allows sideways orientation by specifying the option `horizontal=TRUE`.

6.1.12   Side-by-side boxplots

See also 6.1.11 (boxplots)   *Example:* See 4.7.6 and 5.7.11

```
boxplot(y[x==0], y[x==1], y[x==2], names=c("X=0", "X=1", "X=2")
```

or

```
boxplot(y ~ x)
```

or

```
library(lattice)
bwplot(y ~ x)
```

The `boxplot()` function can be given multiple arguments of vectors to display, or can use a formula interface (which will generate a boxplot for each level of the variable x). A number of useful options are available, including `varwidth` to draw the boxplots with widths proportional to the square root of the number of observations in that group, `horizontal` to reverse the default orientation, `notch` to display notched boxplots, and `names` to specify a vector of labels for the groups. Boxplots can also be created using the `bwplot()` function in `library(lattice)`.

6.1.13   Interaction plots

Example: See 4.7.6

Interaction plots are used to display means by two variables (as in a two-way analysis of variance, 4.1.8).

```
interaction.plot(x1, x2, y)
```

The default statistic to compute is the mean; other options can be specified using the `fun` option.

6.1.14   Plots for categorical data

A variety of less traditional plots can be used to graphically represent categorical data. While these tend to have a low data to ink ratio, they can be useful in figures with repeated multiples [71].

```
mosaicplot(table(x, y, z))
assocplot(table(x, y))
```

The `mosaicplot()` function provides a graphical representations of a two-dimensional or higher contingency table, with the area of each box representing the number of observations in that cell. The `assocplot()` function can be used to display the deviations from independence for a two-dimensional contingency table. Positive deviations of observed minus expected counts are above the line and colored black, while negative deviations are below the line and colored red.

6.1.15   3-D plots

Perspective or surface plots, needle plots, and contour plots can be used to visualize data in three dimensions. These are particularly useful when a response is observed over a grid of two-dimensional values.

```
persp(x, y, z)
contour(x, y, z)
image(x, y, z)

library(scatterplot3d)
scatterplot3d(x, y, z)
```

The values provided for x and y must be in ascending order.

6.1.16   Circular plot

Circular plots are used to analyze data that wrap (e.g., directions expressed as angles, time of day on a 24-hour clock) [16, 33].

```
library(circular)
plot.circular(x, stack=TRUE, bins=50)
```

6.1.17 Receiver operating characteristic (ROC) curve

Example: See 6.6.7

See also 3.3.2 (diagnostic agreement) and 5.1.1 (logistic regression)

Receiver operating characteristic curves can be used to help determine the optimal cut-score to predict a dichotomous measure. This is particularly useful in assessing diagnostic accuracy in terms of sensitivity (the probability of detecting the disorder if it is present), specificity (the probability that a disorder is not detected if it is not present), and the area under the curve (AUC). The variable x represents a predictor (e.g., individual scores) and y a dichotomous outcome. There is a close connection between the idea of the ROC curve and goodness of fit for logistic regression, where the latter allows multiple predictors to be used.

```
library(ROCR)
pred = prediction(x, y)
perf = performance(pred, "tpr", "fpr")
plot(perf)
```

The AUC can be calculated by specifying "auc" as an argument when calling the performance() function.

6.1.18 Kaplan–Meier plot

See also 3.4.4 (logrank test)   *Example:* See 6.6.6

```
library(survival)
fit = survfit(Surv(time, status) ~ as.factor(x), data=ds)
plot(fit, conf.int=FALSE, lty=1:length(unique(x)))
```

The Surv() function is used to combine survival time and status, where time is length of follow-up (interval censored data can be accommodated via an additional parameter) and status=1 indicates an event (e.g., death) while status=0 indicated censoring. A stratified model by each level of the group variable x (see also adding legends, 6.2.15 and different line styles, 6.3.9). More information can be found in the CRAN Survival Task View.

6.1.19   Plot an arbitrary function

Example: See 7.3.4

```
curve(expr, from, to)
```

The `curve()` function can be used to plot an arbitrary function denoted by `expr`, with x values over a specified range (see also 6.2.5, lines).

6.1.20   Empirical cumulative probability density plot

Cumulative density plots are nonparametric estimates of the empirical cumulative probability density function.

```
plot(ecdf(x))
```

6.1.21   Empirical probability density plot

Example: See 3.6.4 and 4.7.4

Density plots are nonparametric estimates of the empirical probability density function.

```
# univariate density
plot(density(x))
```

or

```
library(GenKern)
# bivariate density
op = KernSur(x, y, na.rm=TRUE)
image(op$xords, op$yords, op$zden, col=gray.colors(100),
    axes=TRUE, xlab="x var", ylab="y var"))
```

The bandwidth for `density()` can be specified using the `bw` and `adjust` options, while the default smoother can be specified using the `kernel` option (possible values include the default gaussian, rectangular, triangular, epanechnikov, biweight, cosine, or optcosine). Bivariate density support is provided with the `GenKern` library. Any of the three-dimensional plotting routines (see 6.1.15) can be used to visualize the results.

6.1.22   Normal quantile-quantile plot

Example: See 4.7.4

Quantile-quantile plots are a commonly used graphical technique to assess whether a univariate sample of random variables is consistent with a Gaussian (normal) distribution.

```
qqnorm(x)
qqline(x)
```

The `qqline()` function adds a straight line which goes through the first and third quartiles.

6.2 Adding elements

It is relatively simple to add features to graphs which have been generated by one of the functions discussed in Section 6.1. For example adding an arbitrary line to a graphic requires only one function call (see 6.2.4).

6.2.1 Plot symbols

Example: See 3.6.2

```
plot(x, y, pch=pchval)
```

or

```
points(x, y, string, pch=pchval)
```

or

```
library(lattice)
xyplot(x ~ y, group=factor(groupvar), data=ds)
```

or

```
library(ggplot2)
qplot(x, y, col=factor(groupvar), shape=factor(groupvar),
    data=ds)
```

The `pch` option requires either a single character or an integer code. Some useful values include 20 (dot), 46 (point), 3 (plus), 5 (diamond), and 2 (triangle) (running `example(points)` will display more possibilities). The size of the plotting symbol can be changed using the `cex` option. The vector function `text()` adds the value in the variable `string` to the plot at the specified location. The examples using `xyplot()` and `qplot()` will also generate scatterplots with different plot symbols for each level of `groupvar`.

6.2.2 Add points to an existing graphic

See also 6.2.1 (specifying plotting character)   *Example:* See 4.7.1

```
plot(x, y)
points(x, y)
```

6.2.3   Jitter points

Example: See 3.6.2

Jittering is the process of adding a negligible amount of uniform mean zero noise to each observed value so that the number of observations sharing a value can be easily discerned.

```
jitterx = jitter(x)
```

The default value for the range of the random uniforms is 40% of the smallest difference between values.

6.2.4   Arbitrary straight line

Example: See 4.7.1

```
plot(x, y)
lines(point1, point2)
```

or

```
abline(intercept, slope)
```

The `lines()` function draws a line between the points specified by `point1` and `point2`, which are each vectors with values for the `x` and `y` axes. The `abline()` function draws a line based on the slope-intercept form. Vertical or horizontal lines can be specified using the `v` or `h` option to `abline()`.

6.2.5   OLS line fit to points

See also 4.5.4 (confidence intervals for predicted mean)   *Example:* See 6.6.8

```
plot(x, y)
abline(lm(y ~ x))
```

The `abline()` function accepts regression objects with a single predictor as input.

6.2.6   Smoothed line

See also 5.7.9 (generalized additive models)   *Example:* See 3.6.2

```
plot(...)
lines(lowess(x, y))
```

The `f` parameter to `lowess()` can be specified to control the proportion of points which influence the local value (larger values give more smoothness). The `supsmu()` (Friedman's "super smoother") and `loess()` (local polynomial regression fitting) functions are alternative smoothers.

6.2.7   Add grid

Example: See 2.13.7

A rectangular grid can be added to an existing plot.

```
grid(nx=num, ny=num)
```

The `nx` and `ny` options control the number of cells in the grid. If there are specified as `NULL`, the grid aligns with the tick marks. The default shading is light gray, with a dotted line. Further support for complex grids is available within the `grid.lines()` function in the `grid` package.

6.2.8   Normal density

Example: See 4.7.4

A normal density plot can be added as an annotation to a histogram or empirical density.

```
hist(x)
xvals = seq(from=min(x), to=max(x), length=100)
lines(pnorm(xvals, mean(x), sd(x))
```

6.2.9   Marginal rug plot

Example: See 3.6.2

A rug plot displays the marginal distribution on one of the margins of a scatterplot.

```
rug(x, side=sideval)
```

The `rug()` function adds a marginal plot to one of the sides of an existing plot (`sideval=1` for bottom (default), 2 for left, 3 for top and 4 for right side).

6.2.10   Titles

Example: See 3.6.4

```
title(main="main", sub="sub", xlab="xlab", ylab="ylab")
```

The title commands refer to the main title, subtitle, x-axis, and y-axis, respectively. Some plotting commands (e.g., `hist()`) create titles by default, and the appropriate option within those routines needs to be specified when calling them.

6.2.11   Footnotes

```
title(sub="sub")
```

The `sub` option for the `title()` function generates a subtitle.

6.2.12   Text

Example: See 3.6.2 and 7.4.3

```
text(x, y, labels)
```

Each value of the character vector `labels` is displayed at the specified (X,Y) coordinate. The `adj` option can be used to change text justification to the left, center (default), or right of the coordinate. The `srt` option can be used to rotate text, while `cex` controls the size of the text. The `font` option to `par()` allows specification of plain, bold, italic, or bold italic fonts (see the `family` option to specify the name of a different font family).

6.2.13   Mathematical symbols

Example: See 2.13.7

```
plot(x, y)
text(x, y, expression(mathexpression))
```

The `expression()` argument can be used to embed mathematical expressions and symbols (e.g., $\mu = 0$, $\sigma^2 = 4$) in graphical displays as text, axis labels, legends, or titles. See `help(plotmath)` for more details on the form of `mathexpression` and `example(plotmath)` for examples.

6.2.14   Arrows and shapes

Example: See 3.6.4, 6.1.3, and 6.6.8

```
arrows(x, y)
rect(xleft, ybottom, xright, ytop)
polygon(x, y)
symbols(x, y, type)

library(plotrix)
draw.circle(x, y, r)
```

The `arrows`, `rect()` and `polygon()` functions take vectors as arguments and create the appropriate object with vertices specified by each element of those vectors. Possible `type` values for the `symbols` command include circles, squares, stars, thermometers, and boxplots (see also `library(ellipse)`).

6.2.15 Legend

Example: See 2.13.7 and 3.6.4

```
plot(x, y)
legend(xval, yval, legend=c("Grp1","Grp2"), lty=1:2, col=3:4)
```

The `legend()` command can be used to add a legend at the location (`xval`, `yval`) to distinguish groups on a display. Line styles (6.3.9) and colors (6.3.11) can be used to distinguish the groups. A vector of legend labels, line types, and colors can be specified using the `legend`, `lty`, and `col` options, respectively.

6.2.16 Identifying and locating points

```
locator(n)
```

or

```
identify(x, y, z)
```

The `locator()` function identifies the position of the cursor when the mouse button is pressed. An optional argument `n` specifies how many values to return. The `identify()` function works in the same fashion, but displays the value of the variable `z` that is closest to the selected point.

6.3 Options and parameters

Many options can be given to plots. Depending on the particular situation, these are given as arguments to `plot()`, `par()`, or other high-level functions. Many of these options are described in the documentation for the `par()` function.

6.3.1 Graph size

```
pdf("filename.pdf", width=Xin, height=Yin)
```

The graph size is specified as an optional argument when starting a graphics device (e.g., pdf(), Section 6.4.1). Arguments `Xin` and `Yin` given in inches for `pdf()`, and in pixels for other devices (see the `units` option in `help(jpeg)`).

6.3.2 Point and text size

Example: See 4.7.6

```
plot(x, y, cex=cexval)
```

The `cex` options specified how much the plotting text and symbols should be magnified relative to the default value of 1 (see `help(par)` for details on how to specify this for the axis, labels, and titles, e.g., `cex.axis`).

6.3.3   Box around plots

Example: See 3.6.4

```
plot(x, y, bty=btyval)
```

Control for the box around the plot can be specified using `btyval`, where if the argument is one of `o` (the default), `1`, `7`, `c`, `u`, or `]`, the resulting box resembles the corresponding character, while a value of `n` suppresses the box.

6.3.4   Size of margins

Example: See 4.7.4

The margin options control how tight plots are to the printable area.

```
par(mar=c(bot, left, top, right),    # inner margin
    oma=c(bot, left, top, right))    # outer margin
```

The vector given to `mar` specifies the number of lines of margin around a plot: the default is `c(5, 4, 4, 2) + 0.1`. The `oma` option specifies additional lines outside the entire plotting area (the default is `c(0,0,0,0)`). Other options to control margin spacing include `omd` and `omi`.

6.3.5   Graphical settings

Example: See 4.7.4

```
# change values, while saving old
oldvalues = par(...)


# restore old values for graphical settings
par(oldvalues)
```

6.3.6   Multiple plots per page

Example: See 4.7.4 and 6.6.4

```
par(mfrow=c(a, b))
```

or

```
par(mfcol=c(a, b))
```

The `mfrow` option specifies that plots will be drawn in an a × b array by row (by column for `mfcol`). More complex ways of dividing the graphics device are available through use of the `layout()` function (see also `split.screen()`).

6.3.7 Axis range and style

Example: See 4.7.1 and 6.6.1

```
plot(x, y, xlim=c(minx, maxx), ylim=c(miny, maxy), xaxs="i",
    yaxs="i")
```

The `xaxs` and `yaxs` options control whether tick marks extend beyond the limits of the plotted observations (default) or are constrained to be internal ("i"). More control is available through the `axis()` and `mtext()` functions.

6.3.8 Axis labels, values, and tick marks

Example: See 2.13.7

```
plot(x, y, lab=c(x, y, len),   # number of tick marks
    las=lasval,     # orientation of tick marks
    tck=tckval,     # length of tick marks
    tcl=tclval,     # length of tick marks
    xaxp=c(x1, x2, n),   # coordinates of the extreme tick marks
    yaxp=c(x1, x2, n),   # coordinates of the extreme tick marks
    xlab="X axis label", ylab="Y axis label")
```

Options for `las` include 0 for always parallel, 1 for always horizontal, 2 for perpendicular, and 3 for vertical.

6.3.9 Line styles

Example: See 4.7.4

```
plot(...)
lines(x, y, lty=ltyval)
```

Supported line type values include 0=blank, 1=solid (default), 2=dashed, 3=dotted, 4=dotdash, 5=longdash, and 6=twodash.

6.3.10 Line widths

Example: See 2.13.7

```
plot(...)
lines(x, y, lwd=lwdval)
```

The default for `lwd` is 1; the value of `lwdval` must be positive.

6.3.11   Colors

Example: See 3.6.4

```
plot(...)
lines(x, y, col=colval)
```

For more information on setting colors, see the `Color Specification` section within `help(par)`. The `colors()` function lists available colors, while `colors.plot()` function within the `epitools` package displays a matrix of colors, and `colors.matrix()` returns a matrix of color names. Assistance in finding an appropriate color palette is available from the `display.brewer.all()` function within the `RColorBrewer` package.

6.3.12   Log scale

```
plot(x, y, log=logval)
```

A natural log scale can be specified using the `log` option to `plot()`, where `log="x"` denotes only the x axis, `"y"` only the y axis, and `"xy"` for both.

6.3.13   Omit axes

Example: See 7.4.1

```
plot(x, y, xaxt="n", yaxt="n")
```

6.4   Saving graphs

It is straightforward to export graphics in a variety of formats.

6.4.1   PDF

Example: See 7.4.3

```
pdf("file.pdf")
plot(...)
dev.off()
```

The `dev.off()` function is used to close a graphics device.

6.4.2 Postscript

```
postscript("file.ps")
plot(...)
dev.off()
```

The `dev.off()` function is used to close a graphics device.

6.4.3 JPEG

```
jpeg("filename.jpg")
plot(...)
dev.off()
```

The `dev.off()` function is used to close a graphics device.

6.4.4 WMF

```
win.metafile("file.wmf")
plot(...)
dev.off()
```

The function `win.metafile()` is only supported under Windows. Functions which generate multiple plots are not supported. The `dev.off()` function is used to close a graphics device.

6.4.5 BMP

```
bmp("filename.bmp")
plot(...)
dev.off()
```

The `dev.off()` function is used to close a graphics device.

6.4.6 TIFF

```
tiff("filename.tiff")
plot(...)
dev.off()
```

The `dev.off()` function is used to close a graphics device.

6.4.7 PNG

```
png("filename.png")
plot(...)
dev.off()
```

The `dev.off()` function is used to close a graphics device.

6.4.8 Closing a graphic device

Example: See 6.6.6

The `dev.off()` function closes a graphics window. This is particularly useful when a graphics file is being created.

```
dev.off()
```

6.5 Further resources

The books by Tufte [70, 71, 72, 73] provide an excellent framework for graphical displays, some of which build on the work of Tukey [74]. Comprehensive and accessible books on R graphics include the texts by Murrell [45] and Sarkar [61].

6.6 HELP examples

To help illustrate the tools presented in this chapter, we apply many of the entries to the HELP data. The code for these examples can be downloaded from http://www.math.smith.edu/r/examples.

```
> options(digits=3)
> library(foreign)
> ds = read.dta("http://www.math.smith.edu/r/data/help.dta")
> attach(ds)
```

6.6.1 Scatterplot with multiple axes

The following example creates a single figure that displays the relationship between CESD and the variables `indtot` (Inventory of Drug Use Consequences, InDUC) and `mcs` (Mental Component Score), for a subset of female alcohol-involved subjects. We specify two different y-axes (6.1.2) for the figure.

A nontrivial amount of housekeeping is needed. The second y variable must be rescaled to the range of the original, and the axis labels and tick marks added on the right. To accomplish this, we write a function `plottwoy()` which first makes the plot of the first (left axis) y against x, adds a lowess curve through that data, then calls a second function, `addsecondy()`.

```
> plottwoy = function(x, y1, y2, xname="X", y1name="Y1",
+    y2name="Y2")
+ {
+    plot(x, y1, ylab=y1name, xlab=xname)
+    lines(lowess(x, y1), lwd=3)
+    addsecondy(x, y2, y1, yname=y2name)
+ }
```

The function `addsecondy()` does the work of rescaling the range of the second variable to that of the first, adds the right axis, and plots a lowess curve through the data for the rescaled y2 variable.

```
> addsecondy = function(x, y, origy, yname="Y2") {
+    prevlimits = range(origy)
+    axislimits = range(y)
+    axis(side=4, at=prevlimits[1] + diff(prevlimits)*c(0:5)/5,
+       labels=round(axislimits[1] + diff(axislimits)*c(0:5)/5,
+       1))
+    mtext(yname, side=4)
+    newy = (y-axislimits[1])/
+       (diff(axislimits)/diff(prevlimits)) + prevlimits[1]
+    points(x, newy, pch=2)
+    lines(lowess(x, newy), lty=2, lwd=3)
+ }
```

Finally, the newly defined functions can be run and Figure 6.1 generated.

```
> plottwoy(cesd[female==1&substance=="alcohol"],
+    indtot[female==1&substance=="alcohol"],
+    mcs[female==1&substance=="alcohol"], xname="cesd",
+    y1name="InDUC", y2name="mcs")
```

6.6.2 Bubble plot

Figure 6.2 displays a bubble plot (6.1.3) using circles as the plotting symbol (6.2.14). Amongst female subjects with alcohol as a primary substance, the circles are plotted by age and CESD score, with the area of the circles proportional to the number of drinks.

```
> femalealc = subset(ds, female==1 & substance=="alcohol")
> with(femalealc, symbols(age, cesd, circles=sqrt(i1),
+    inches=1/5, bg=ifelse(homeless, "lightgray", "white")))
```

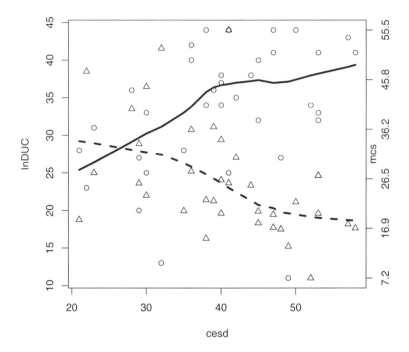

Figure 6.1: Plot of InDUC and MCS versus CESD for female alcohol-involved subjects.

The homeless subjects tend to have higher CESD scores than non-homeless subjects, and the rate of drinking also appears to be positively associated with CESD scores.

6.6.3   Conditioning plot

Figure 6.3 displays a conditioning plot (6.1.6) with the association between MCS and CESD stratified by substance and report of suicidal thoughts (g1b).

  We need to ensure that the necessary packages are installed (1.7.1) then set up and generate the plot.

```
> library(lattice)
> suicidal.thoughts = as.factor(g1b)
> coplot(mcs ~ cesd | suicidal.thoughts*substance,
+    panel=panel.smooth)
```

There is a similar association between CESD and MCS for each of the substance groups. Subjects with suicidal thoughts tended to have higher CESD scores,

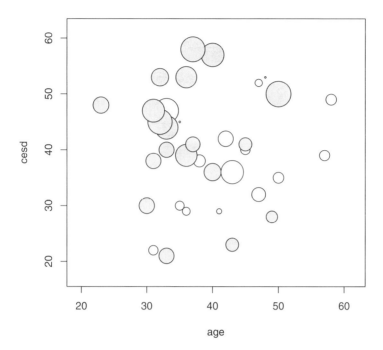

Figure 6.2: Distribution of number of drinks (proportional to area of circle), age, and CESD for female alcohol-involved subjects (homeless subjects are shaded).

and the association between CESD and MCS was somewhat less pronounced than for those without suicidal thoughts.

6.6.4 Multiple plots

Figure 6.4 displays a graphic built up using multiple plots (6.3.6) with the layout() function. The association between MCS and CESD is stratified by gender, with histograms on the margins. The code divides the plotting area into rows and columns with different widths, then adds each graphic element.

```
> xhist = hist(mcs, plot=FALSE)
> yhist = hist(cesd, plot=FALSE)
> top = max(c(xhist$counts, yhist$counts))
> nf = layout(matrix(c(1,2,3,4),2,2,byrow=TRUE), widths=c(4,1),
+     heights=c(1,4), TRUE)
```

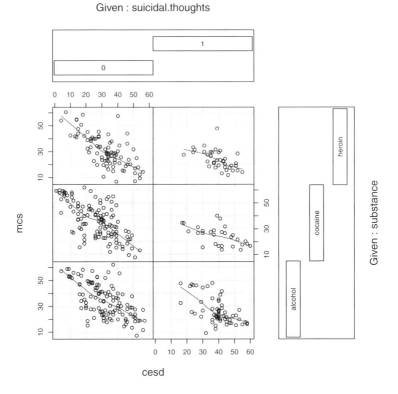

Figure 6.3: Association of MCS and CESD, stratified by substance and report of suicidal thoughts.

```
> par(mar=c(0,3,1,1)) # top histogram
> barplot(xhist$counts, axes=FALSE, ylim=c(0, top), space=0)
> par(mar=c(4,4,1,1))  # empty plot
> plot(mcs, cesd, type="n", xlab="", ylab="", xaxt="n", yaxt="n",
+    bty="n")
> par(mar=c(4,4,1,1)) # main scatterplot
> plot(mcs, cesd, xlab="", ylab="",
+    pch=ifelse(female, "F", "M"), cex=.6)
> lines(lowess(mcs[female==1], cesd[female==1]), lwd=2)
> lines(lowess(mcs[female==0], cesd[female==0]), lwd=3, lty=4)
> text(35, max(cesd),"MCS")
> text(max(mcs), 30,"CESD", srt=270)
> legend(min(mcs), 10, legend=c("Female","Male"), lwd=3,
+    lty=c(1,4))
> par(mar=c(3,0,1,1)) # rightside histogram
> barplot(yhist$counts, axes=FALSE, xlim=c(0, top), space=0,
+    horiz=TRUE)
```

There is a similar association between CESD and MCS for men and women, though women tend to have higher CESD scores.

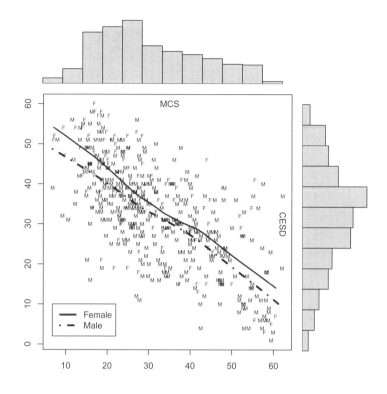

Figure 6.4: Association of MCS and CESD, stratified by gender.

6.6.5 Dotplot

There is an association between CESD scores and the SF-36 mental (MCS) and physical (PCS) scores. These can be displayed using a dotplot (6.1.8), where the average MCS and PCS score is calculated for each value of CESD score. To help smooth the display, odd values of CESD scores are combined with the next lowest even number (e.g., the values of MCS and PCS for CESD scores of both 0 and 1 are averaged).

```
> cesdeven = cesd - cesd %% 2      # map to lowest even number
> pcsvals = tapply(pcs, cesdeven, mean)   # calculate averages
> mcsvals = tapply(mcs, cesdeven, mean)
> library(lattice)
> print(dotplot( ~ mcsvals + pcsvals, xlab="MCS/PCS scores",
+    ylab="CESD score", pch=c("M", "P"), cex=1.4, col=c(1,1)))
```

Figure 6.5 displays the means, with separate plotting characters for MCS and
PCS. There is a stronger association between CESD and MCS than between
CESD and PCS.

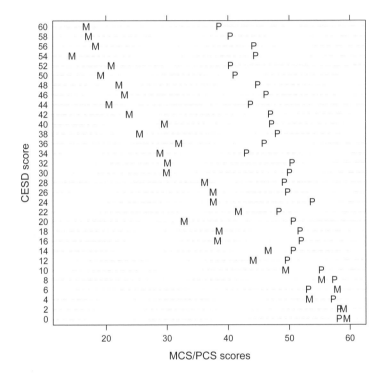

Figure 6.5: Mean MCS and PCS scores by CESD.

6.6.6   Kaplan–Meier plot

The main outcome of the HELP study was time to linkage to primary care, as a
function of randomization group. This can be displayed using a Kaplan–Meier
plot (see 6.1.18). Detailed information regarding the Kaplan–Meier estimator

at each time point can be found by specifying summary(survobj). Figure 6.6 displays the estimates, with + signs indicating censored observations.

```
> library(survival)
> survobj = survfit(Surv(dayslink, linkstatus) ~ treat)
> print(survobj)

          records n.max n.start events median 0.95LCL 0.95UCL
treat=0       209   209     209     35     NA      NA      NA
treat=1       222   222     222    128    120      79     272

> plot(survobj, lty=1:2, lwd=2, col=c(4,2), conf.int=TRUE,
+     xlab="days", ylab="P(not linked)")
> title("Product-Limit Survival Estimates")
> legend(230, .75, legend=c("Control", "Treatment"), lty=c(1,2),
+     lwd=2, col=c(4,2), cex=1.4)
```

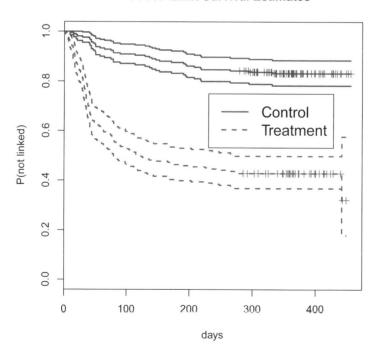

Figure 6.6: Kaplan–Meier estimate of time to linkage to primary care by randomization group.

As reported previously by Horton et al. and Samet et al. [29, 59], there is a highly statistically significant effect of treatment, with approximately 55% of clinic subjects linking to primary care, as opposed to only 15% of control subjects.

6.6.7   ROC curve

Receiver operating characteristic (ROC) curves are used for diagnostic agreement (3.3.2 and 6.1.17) as well as assessing goodness of fit for logistic regression (5.1.1). These can be generated using the ROCR library.

Using R, we first create a prediction object, then retrieve the area under the curve (AUC) to use in Figure 6.7.

Figure 6.7 displays the receiver operating characteristic curve predicting suicidal thoughts using the CESD measure of depressive symptoms.

```
> library(ROCR)
> pred = prediction(cesd, g1b)
> auc = slot(performance(pred, "auc"), "y.values")[[1]]
```

We can then plot the ROC curve, adding display of cutoffs for particular CESD values ranging from 20 to 50. These values are offset from the ROC curve using the text.adj option.

If the continuous variable (in this case cesd) is replaced by the predicted probability from a logistic regression model, multiple predictors can be included.

```
> plot(performance(pred, "tpr", "fpr"),
+      print.cutoffs.at=seq(from=20, to=50, by=5),
+      text.adj=c(1, -.5), lwd=2)
> lines(c(0, 1), c(0, 1))
> text(.6, .2, paste("AUC=", round(auc,3), sep=""), cex=1.4)
> title("ROC Curve for Model")
```

6.6.8   Pairs plot

We can qualitatively assess the associations between some of the continuous measures of mental health, physical health, and alcohol consumption using a pairsplot or scatterplot matrix (6.1.5). To make the results clearer, we display only the female subjects.

A simple version with only the scatterplots could be generated easily with the pairs() function (results not shown):

```
> pairs(c(ds[72:74], ds[67]))
```

or

```
> pairs(ds[c("pcs", "mcs", "cesd", "i1")])
```

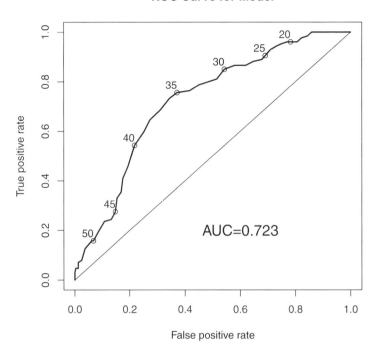

Figure 6.7: Receiver operating characteristic curve for the logistical regression model predicting suicidal thoughts using the CESD as a measure of depressive symptoms (sensitivity = true positive rate; 1-specificity = false positive rate).

Here instead we demonstrate building a figure using several functions. We begin with a function `panel.hist()` to display the diagonal entries (in this case, by displaying a histogram).

```
> panel.hist = function(x, ...)
+ {
+     usr = par("usr"); on.exit(par(usr))
+     par(usr = c(usr[1:2], 0, 1.5) )
+     h = hist(x, plot=FALSE)
+     breaks = h$breaks; nB = length(breaks)
+     y = h$counts; y = y/max(y)
+     rect(breaks[-nB], 0, breaks[-1], y, col="cyan", ...)
+ }
```

Another function is created to create a scatterplot along with a fitted line.

```
> panel.lm = function(x, y, col=par("col"), bg=NA,
+    pch=par("pch"), cex=1, col.lm="red", ...)
+ {
+    points(x, y, pch=pch, col=col, bg=bg, cex=cex)
+    ok = is.finite(x) & is.finite(y)
+    if (any(ok))
+        abline(lm(y[ok] ~ x[ok]))
+ }
```

The panel.lm() function uses indexing (1.5.2) to prune infinite values, and
then add a line (6.2.4) based on this subset. These functions are called (along
with the built-in panel.smooth() function) to display the results. Figure 6.8
displays the pairsplot of CESD, MCS, PCS, and I1, with histograms along the
diagonals. Smoothing splines are fit on the lower triangle, linear fits on the
upper triangle, using code fragments derived from example(pairs).

```
> pairs(~ cesd + mcs + pcs + i1, subset=(female==1),
+    lower.panel=panel.smooth, diag.panel=panel.hist,
+    upper.panel=panel.lm)
```

There is an indication that CESD, MCS, and PCS are interrelated, while I1 appears
to have modest associations with the other variables.

6.6.9   Visualize correlation matrix

One visual analysis which might be helpful to display would be the pairwise
correlations for a set of variables. We utilize the approach used by Sarkar
to recreate Figure 13.5 of the *Lattice: Multivariate data visualization with R*
book [61]. Other examples in that reference help to motivate the power of the
lattice package far beyond what is provided by demo(lattice).

```
> cormat = cor(cbind(mcs, pcs, pss_fr, drugrisk, cesd, indtot,
+    i1, sexrisk), use="pairwise.complete.obs")
> oldopt = options(digits=1)
```

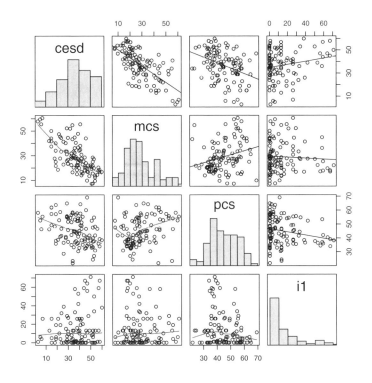

Figure 6.8: Pairsplot of variables from the HELP dataset.

```
> cormat

          mcs   pcs pss_fr drugrisk  cesd indtot    i1 sexrisk
mcs      1.00  0.11   0.14   -0.206 -0.68   -0.4 -0.09  -0.106
pcs      0.11  1.00   0.08   -0.141 -0.29   -0.1 -0.20   0.024
pss_fr   0.14  0.08   1.00   -0.039 -0.18   -0.2 -0.07  -0.113
drugrisk -0.21 -0.14  -0.04    1.000  0.18    0.2 -0.10  -0.006
cesd     -0.68 -0.29  -0.18    0.179  1.00    0.3  0.18   0.016
indtot   -0.38 -0.13  -0.20    0.181  0.34    1.0  0.20   0.113
i1       -0.09 -0.20  -0.07   -0.100  0.18    0.2  1.00   0.088
sexrisk  -0.11  0.02  -0.11   -0.006  0.02    0.1  0.09   1.000

> options(oldopt)    # return options to previous setting
```

```
> drugrisk[is.na(drugrisk)] = 0
> panel.corrgram = function(x, y, z, at, level=0.9,
+    label=FALSE, ...)
+ {
+    require("ellipse", quietly=TRUE)
+    zcol = level.colors(z, at=at, col.regions=gray.colors)
+    for (i in seq(along=z)) {
+        ell = ellipse(z[i], level=level, npoints=50,
+            scale=c(.2, .2), centre=c(x[i], y[i]))
+        panel.polygon(ell, col=zcol[i], border=zcol[i], ...)
+    }
+    if (label)
+        panel.text(x=x, y=y, lab=100*round(z, 2), cex=0.8,
+            col=ifelse(z < 0, "white", "black"))
+ }
```

The `panel.corrgram()` function uses the `ellipse` package to display the correlations, with depth of shading proportional to the magnitude of the correlation. Negative associations are marked using a white font, while positive associations are displayed using a black font.

```
> library(ellipse)
> library(lattice)
> print(levelplot(cormat, at=do.breaks(c(-1.01, 1.01), 20),
+    xlab=NULL, ylab=NULL, colorkey=list(space = "top",
+    col=gray.colors), scales=list(x=list(rot = 90)),
+    panel=panel.corrgram,
+    label=TRUE))
```

Figure 6.9 displays the results, including a color key at the top of the graph.

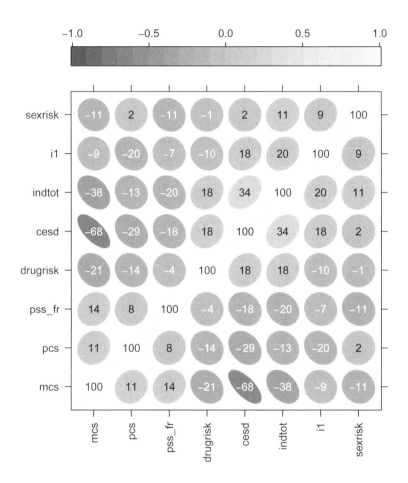

Figure 6.9: Visual display of associations.

Chapter 7

Advanced applications

In this chapter, we address several additional topics that show off the statistical computing strengths and potential of R, as well as illustrate many of the entries in the earlier chapters.

7.1 Power and sample size calculations

Many simple settings lend themselves to analytic power calculations, where closed-form solutions are available. Other situations may require an empirical calculation, where repeated simulation is undertaken. In addition to the examples in this section other power routines are available in `library(pwr)` and `library(Hmisc)`.

7.1.1 Analytic power calculation

It is straightforward to find power (for a given sample size) or sample size (given a desired power) for two-sample comparisons of either continuous or categorical outcomes. We show simple examples for comparing means and proportions in two groups.

```
# find sample size for two-sample t-test
power.t.test(delta=0.5, power=0.9)
```

```
# find power for two-sample t-test
power.t.test(delta=0.5, n=100)
```

The latter call generates the following output.

```
    Two-sample t test power calculation
              n = 100
          delta = 0.5
             sd = 1
      sig.level = 0.05
          power = 0.9404272
    alternative = two.sided
 NOTE: n is number in *each* group
```

```
# find sample size for two-sample test of proportions
power.prop.test(p1=.1, p2=.2, power=.9)
```

```
# find power for two-sample test of proportions
power.prop.test(p1=.1, p2=.2, n=100)
```

The power.t.test() function requires exactly four of the five arguments (sample size in each group, power, difference between groups, standard deviation, and significance level) to be specified. Default values are set to sig.level=0.05 and sd=1.

7.1.2   Simulation-based power calculations

In some settings, analytic power calculations may not be readily available. A straightforward alternative is to estimate power empirically, simulating data from the proposed design under various alternatives.

We consider a study of children clustered within families. Each family has 3 children; in some families all 3 children have an exposure of interest, while in others just 1 child is exposed. In the simulation, we assume that the outcome is multivariate normal with higher mean for those with the exposure, and 0 for those without. A compound symmetry correlation is assumed, with equal variances at all times. We assess the power to detect an exposure effect where the intended analysis uses a random intercept model (5.2.2) to account for the clustering within families.

With this simple covariance structure it is trivial to generate correlated errors. We specify the correlation matrix directly, and simulate from a multivariate normal distribution.

```
library(MASS)
library(nlme)

# initialize parameters and building blocks

# effect size
effect = 0.35

# intrafamilial correlation
corr = 0.4
numsim = 1000

# families with 3 exposed
n1fam = 50

# families with 1 exposed and 2 unexposed
n2fam = 50

# 3x3 compound symmetry correlation
vmat = matrix(c
    (1,    corr, corr,
    corr, 1   , corr,
    corr, corr, 1    ), 3, 3)

# create exposure status for all sets of families
# 1 1 1 ... 1 0 0 0 ... 0
x = c(rep(1, n1fam), rep(1, n1fam), rep(1, n1fam),
     rep(1, n2fam), rep(0, n2fam), rep(0, n2fam))

# create identifiers for families
# 1 2 ... n1fam 1 2 ... n1fam ...
id = c(1:n1fam, 1:n1fam, 1:n1fam,
    (n1fam+1:n2fam), (n1fam+1:n2fam), (n1fam+1:n2fam))

# initialize vector for results
power = numeric(numsim)
```

The concatenate function (c(), Section 1.5.1) is used to glue together the appropriate elements of the design matrices and underlying correlation structure.

```
for (i in 1:numsim) {
    cat(i," ")
    # all three exposed
    grp1 = mvrnorm(n1fam, c(effect, effect, effect), vmat)

    # only first exposed
    grp2 = mvrnorm(n2fam, c(effect, 0,        0),        vmat)

    # concatenate the outcome vector
    y = c(grp1[,1], grp1[,2], grp1[,3],
          grp2[,1], grp2[,2], grp2[,3])

    # specify dependence structure
    group = groupedData(y ~ x | id)
    # fit random intercept model
    res = lme(group, random = ~ 1)
    # grab results for main parameter
    pval = summary(res)$tTable[2,5]
    # is it statistically significant?
    power[i] = pval<=0.05
}
cat("\nEmpirical power for effect size of ", effect,
    " is ", round(sum(power)/numsim,3), ".\n", sep="")
cat("95% confidence interval is",
    round(prop.test(sum(power), numsim)$conf.int, 3), "\n")
```

These assumptions yield the following estimate of power.

```
Empirical power for effect size of 0.35 is 0.855.
95% confidence interval is 0.831 0.876
```

7.2 Simulations and data generation

7.2.1 Simulate data from a logistic regression

As our first example of data generation, we simulate data from a logistic regression (5.1.1). Our process is to generate the linear predictor, then apply the inverse link, and finally draw from a distribution with this parameter. This approach is useful in that it can easily be applied to other generalized linear models (5.1). Here we assume an intercept of 0, a slope of 0.5, and generate 5,000 observations.

```
intercept = 0
beta = 0.5
n = 5000
xtest = rnorm(n, 1, 1)
linpred = intercept + (xtest * beta)
prob = exp(linpred)/(1 + exp(linpred))
ytest = ifelse(runif(n) < prob, 1, 0)
```

We can display estimated values of the coefficients (4.5.1) from the logistic regression model.

```
> coef(glm(ytest ~ xtest, binomial))
(Intercept)        xtest
      0.019        0.506
```

7.2.2 Simulate data from generalized linear mixed model

In this example, we generate clustered data with a dichotomous outcome, for input to a generalized linear mixed model (5.2.6). In the code below, for 3000 clusters (denoted by id) there is a cluster invariant predictor (X_1), 3 observations within each cluster (denoted by X_2) and a linear effect of order within cluster, and an additional predictor which varies between clusters (X_3). The dichotomous outcome Y is generated from these predictors using a logistic link incorporating a random intercept for each cluster.

```
library(lme4)
n = 3000; p = 3; sigbsq = 4
beta = c(-2, 1.5, 0.5, -1)
id = rep(1:n, each=p)    # 1 1 ... 1 2 2 ... 2 ... n
x1 = as.numeric(id < (n+1)/2)  # 1 1 ... 1 0 0 ... 0
randint = rep(rnorm(n, 0, sqrt(sigbsq)), each=p)
x2 = rep(1:p, n)          # 1 2 ... p 1 2 ... p ...
x3 = runif(p*n)
linpred = beta[1] + beta[2]*x1 + beta[3]*x2 + beta[4]*x3 +
    randint
expit = exp(linpred)/(1 + exp(linpred))
y = runif(p*n) < expit

glmmres = lmer(y ~ x1 + x2 + x3 + (1|id),
    family=binomial(link="logit"))
```

This generates the following output.

```
> summary(glmmres)
Generalized linear mixed model fit by the Laplace approximation
Formula: y ~ x1 + x2 + x3 + (1 | id)
   AIC   BIC logLik deviance
 10637 10672  -5313    10627
Random effects:
 Groups Name        Variance Std.Dev.
 id     (Intercept) 3.05     1.75
Number of obs: 9000, groups: id, 3000

Fixed effects:
            Estimate Std. Error z value Pr(>|z|)
(Intercept) -1.8614     0.1021   -18.2  <2e-16
x1           1.3827     0.0838    16.5  <2e-16
x2           0.4903     0.0323    15.2  <2e-16
x3          -1.0338     0.1014   -10.2  <2e-16
---
Correlation of Fixed Effects:
   (Intr) x1     x2
x1 -0.452
x2 -0.651  0.047
x3 -0.443 -0.024 -0.041
```

7.2.3 Generate correlated binary data

Correlated dichotomous outcomes Y_1 and Y_2 with a desired association and success probability can be generated by finding the probabilities corresponding to the 2×2 table as a function of the marginal expectations and correlation using the methods of Lipsitz and colleagues [41]. Here we generate a sample of 1,000 values, where: $P(Y_1 = 1) = 0.15$, $P(Y_2 = 1) = 0.25$ and $\mathrm{Corr}(Y_1, Y_2) = 0.40$.

```
p1 = 0.15; p2 = 0.25; corr = 0.4; n = 10000
p1p2 = corr*sqrt(p1*(1-p1)*p2*(1-p2)) + p1*p2
vals = sample(1:4, n, replace=TRUE,
    prob=c(1-p1-p2+p1p2, p1-p1p2, p2-p1p2, p1p2))
y1 = numeric(n); y2 = y1           # create output vectors
y1[vals==2 | vals==4] = 1          # and replace them with ones
y2[vals==3 | vals==4] = 1          # where needed
rm(vals, p1, p2, p1p2, corr, n)    # cleanup
```

The generated data is similar to the expected values.

```
> cor(y1, y2)
[1] 0.392
> mean(y1)
[1] 0.151
> mean(y2)
[1] 0.248
```

7.2.4 Simulate data from a Cox model

To simulate data from a Cox proportional hazards model (5.3.1), we need to model the hazard functions for both time to event and time to censoring. In this example, we use a constant baseline hazard, but this can be modified by specifying other `scale` parameters for the Weibull random variables.

```
library(survival)
n = 10000
beta1 = 2; beta2 = -1
lambdaT = .002 # baseline hazard
lambdaC = .004  # hazard of censoring

# generate data from Cox model
x1 = rnorm(n,0)
x2 = rnorm(n,0)
# true event time
T = rweibull(n, shape=1, scale=lambdaT*exp(-beta1*x1-beta2*x2))
C = rweibull(n, shape=1, scale=lambdaC)   #censoring time
time = pmin(T,C)  #observed time is min of censored and true
event = time==T   # set to 1 if event is observed

# fit Cox model
survobj = coxph(Surv(time, event)~ x1 + x2, method="breslow")
```

This generates data where approximately 40% of the observations are censored. The results are similar to the true parameter values.

```
> table(event)
event
FALSE   TRUE
 4083   5917
> print(survobj)
Call:
coxph(formula = Surv(time, event) ~ x1 + x2, method = "breslow")

      coef exp(coef) se(coef)      z p
x1  2.02      7.556   0.0224   90.1 0
x2 -1.02      0.362   0.0159  -63.7 0

Likelihood ratio test=11692  on 2 df, p=0   n= 10000
> confint(survobj)
      2.5 %      97.5 %
x1   1.978       2.066
x2  -1.047      -0.985
```

7.3   Data management and related tasks

7.3.1   Finding two closest values in a vector

Suppose we need to find the closest pair of observations for some variable. This might arise if we were concerned that some data had been accidentally duplicated. In this case study, we return the ID's of the two closest observations, and their distance from each other. We will first create some sample data and sort it, recognizing that the smallest difference must come between two adjacent observations, once they are sorted.

We begin by generating data (2.10.6), along with some subject identifiers (2.4.19). The order() function (Section 2.5.6) is used to keep track of the sorted random variables. The ID of the smaller of the two observations with the smallest distance is the value in the id vector in the place where x equals the value of the sorted vector that is in the same place as the smallest difference. The larger ID can be found the same way, using the shifted vector. The which.min()

```
id = 1:10
x = rnorm(10)
sortx = x[order(x)]
oneless = sortx[2:length(x)]
diff = oneless - sortx[1:length(x)-1]
smallid = id[x == sortx[which.min(diff)]]
largeid = id[x == oneless[which.min(diff)]]
smalldist = min(diff)
```

To help clarify this process, a number of relevant intermediate quantities are displayed below.

```
> x
 [1]  1.412 -0.518  1.298  0.351  2.123 -1.388
 [7]  0.431  1.268  0.658 -0.014
> sortx
 [1] -1.388 -0.518 -0.014  0.351  0.431  0.658
 [7]  1.268  1.298  1.412  2.123
> oneless
 [1] -0.518 -0.014  0.351  0.431  0.658  1.268
 [8]  1.298  1.412  2.123
> diff
 [1] 0.87 0.50 0.36 0.08 0.23 0.61 0.03 0.11 0.71
> smallid
 [1] 8
> largeid
 [1] 3
> smalldist
 [1] 0.03
```

7.3.2 Calculate and plot a running average

The "Law of Large Numbers" concerns the convergence of the arithmetic average to the expected value, as sample sizes increase. This is an important topic in mathematical statistics. The convergence (or lack thereof, for certain distributions) can easily be visualized [25]. Assume that X_1, X_2, ..., X_n are independent and identically distributed realizations from some distribution with mean μ. We denote the average of the first k observations as $\bar{X}^{(k)}$.

We define a function (1.6.2) to calculate the running average for a given vector, allowing for variates from many distributions to be generated.

```
runave = function(n, gendist, ...) {
   x = gendist(n, ...)
   avex = numeric(n)
   for (k in 1:n) {
      avex[k] = mean(x[1:k])
   }
   return(data.frame(x, avex))
}
```

The `runave()` function takes at a minimum two arguments: a sample size `n` and function (1.6) denoted by `gendist` that is used to generate samples from a distribution (2.1). In addition, other options for the function can be specified, using the ... syntax (see 1.6). This is used to specify the degrees of freedom for the samples generated for the t distribution in the next code block. The loop in the `runave()` function could be eliminated through use of the `cumsum()` function applied to the vector given as argument, and then divided by a vector of observation numbers.

Next, we generate the data, using our new function. To make sure we have a nice example, we first set a fixed seed (2.10.11). Recall that because the expectation of a Cauchy random variable is undefined [53] the sample average does not converge to the center, while a t distribution with more than 1 df does.

```
vals = 1000
set.seed(1984)
cauchy = runave(vals, rcauchy)
t4 = runave(vals, rt, 4)
```

We now plot the results, beginning with an empty plot with the correct axis limits, using the `type="n"` specification (6.1.1). We add the running average using the lines function (6.2.4) and varying the line style (6.3.9) and thickness (6.3.10) with the `lty` and `lwd` specifications, respectively. Finally we specify a title (6.2.10) and a legend (6.2.15). The results are displayed in Figure 7.1.

```
plot(c(cauchy$avex, t4$avex), xlim=c(1, vals), type="n")
lines(1:vals, cauchy$avex, lty=1, lwd=2)
lines(1:vals, t4$avex, lty=2, lwd=2)
abline(0, 0)
title("Running average of two distributions")
legend(vals*.6, -1, legend=c("Cauchy", "t with 4 df"),
   lwd=2, lty=c(1,2))
```

7.3.3   Tabulate binomial probabilities

Suppose we wanted to assess the probability $P(X = x)$ for a binomial random variate with parameters $n = 10$ and p ranging from $0.81, 0.84, \ldots, 0.99$. This could be helpful, for example, in various game settings or for teaching purposes.

We address this problem by making a vector of the binomial probabilities, using the : operator (2.4.19) to generate a sequence of integers. After creating a matrix (2.9) to hold the table results, we loop (2.11.1) through the binomial probabilities, calling the `dbinom()` function (2.10.2) to find the probability that the random variable takes on that particular value. This calculation is nested within the `round()` function (2.8.4) to reduce the digits displayed. Finally, we glue the vector of binomial probabilities to the results using `cbind()`.

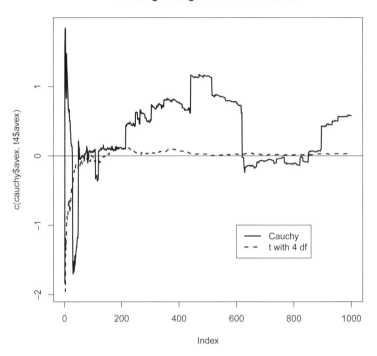

Figure 7.1: Running average for Cauchy and t distributions.

```
p = .78 + (3 * 1:7)/100     # .81 to .99
allprobs = matrix(nrow=length(p), ncol=11)
for (i in 1:length(p)) {
    allprobs[i,] = round(dbinom(0:10, 10, p[i]),2)
}
table = cbind(p, allprobs)
```

This generates the following results.

```
> table
          p
[1,]  0.81 0 0 0 0 0 0.02 0.08 0.19 0.30 0.29 0.12
[2,]  0.84 0 0 0 0 0 0.01 0.05 0.15 0.29 0.33 0.17
[3,]  0.87 0 0 0 0 0 0.00 0.03 0.10 0.25 0.37 0.25
[4,]  0.90 0 0 0 0 0 0.00 0.01 0.06 0.19 0.39 0.35
[5,]  0.93 0 0 0 0 0 0.00 0.00 0.02 0.12 0.36 0.48
[6,]  0.96 0 0 0 0 0 0.00 0.00 0.01 0.05 0.28 0.66
[7,]  0.99 0 0 0 0 0 0.00 0.00 0.00 0.00 0.09 0.90
```

7.3.4   Sampling from a pathological distribution

Evans and Rosenthal [12] consider ways to sample from a distribution with density given by:

$$f(y) = c \exp(-y^4)(1 + |y|)^3,$$

where c is a normalizing constant and y is defined on the whole real line. Use of the probability integral transform (Section 2.10.10) is not feasible in this setting, given the complexity of inverting the cumulative density function.

We can find the normalizing constant c using symbolic mathematics software (e.g., Wolfram Alpha, searching for `int(exp(-y^4)(1+y)^3, y=0..infinity)`). This yielded a result of $\frac{1}{4} + \frac{3\sqrt{\pi}}{4} + \Gamma(5/4) + \Gamma(7/4)$ for the integral over the positive real line, which when doubled gives a value of $c = 6.809610784$.

```
> options(digits=10)
> 2*(1/4 + 3*sqrt(pi)/4 + gamma(5/4) + gamma(7/4))
[1] 6.809610784
```

The Metropolis–Hastings algorithm is a Markov Chain Monte Carlo (MCMC) method for obtaining samples from a probability distribution. The premise for this algorithm is that it chooses proposal probabilities so that after the process has converged we are generating draws from the desired distribution. A further discussion can be found in Section 11.3 of *Probability and Statistics: The Science of Uncertainty* [12] or in Section 1.9 of Gelman et al. [21].

We find the acceptance probability $\alpha(x, y)$ in terms of two densities, $f(y)$ and $q(x, y)$ (a proposal density, in our example, normal with specified mean and unit variance) so that

$$
\begin{aligned}
\alpha(x, y) &= \min\left\{1, \frac{cf(y)q(y, x)}{cf(x)q(x, y)}\right\} \\
&= \min\left\{1, \frac{c\exp(-y^4)(1 + |y|)^3(2\pi)^{-1/2}\exp(-(y-x)^2/2)}{c\exp(-x^4)(1 + |x|)^3(2\pi)^{-1/2}\exp(-(x-y)^2/2)}\right\} \\
&= \min\left\{1, \frac{\exp(-y^4 + x^4)(1 + |y|)^3}{(1 + |x|)^3}\right\}
\end{aligned}
$$

Begin by picking an arbitrary value for X_1. The Metropolis–Hastings algorithm proceeds by computing the value X_{n+1} as follows:
1. Generate y from a Normal(X_n, 1).
2. Compute $\alpha(x, y)$ as above.
3. With probability $\alpha(x, y)$, let $X_{n+1} = y$ (use proposal value). Otherwise, with probability $1 - \alpha(x, y)$, let $X_{n+1} = X_n = x$ (keep previous value).

The code uses the first 50,000 iterations as a burn-in period, then generates 100,000 samples using this procedure (saving every 20th sample to reduce auto-correlation).

```
alphafun = function(x, y) {
    return(exp(-y^4+x^4)*(1+abs(y))^3*
        (1+abs(x))^-3)
}


numvals = 100000; burnin = 50000
i = 1
xn = rnorm(1) # arbitrary value to start
for (i in 1:burnin) {
    propy = rnorm(1, xn, 1)
    alpha = min(1, alphafun(xn, propy))
    xn = sample(c(propy, xn), 1, prob=c(alpha,1-alpha))
}
i = 1
res = numeric(numvals)
while (i <= numvals*20) {
    propy = rnorm(1, xn, 1)
    alpha = min(1, alphafun(xn, propy))
    xn = sample(c(propy, xn), 1, prob=c(alpha,1-alpha))
    if (i%%20==0) res[i/20] = xn
    i = i + 1
}
```

The results are displayed in Figure 7.2, with the dashed line indicating the true distribution, and solid the simulated variates. While the normalizing constant drops out of the algorithm above, it is needed for the plotting done by the curve() function.

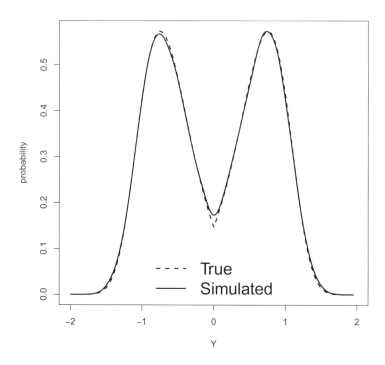

Figure 7.2: Plot of true and simulated distributions.

```
pdfeval = function(x) {
    return(1/6.809610784*exp(-x^4)*(1+abs(x))^3)
}
curve(pdfeval, from=-2, to=2, lwd=2, lty=2, type="l",
    ylab="probability", xlab="Y")
lines(density(res), lwd=2, lty=1)
legend(-1, .1, legend=c("True", "Simulated"),
    lty=2:1, lwd=2, cex=1.8, bty="n")
```

Care is always needed when using MCMC methods. This example was particularly well-behaved, in that the proposal distribution is large compared to the distance between the two modes. Section 6.2 of Lavine [36] and Gelman et al. [21] provides an accessible discussion of these and other issues.

7.4 Read geocoded data and draw maps

7.4.1 Read variable format files and plot maps

Sometimes datasets are stored in variable format. For example, U.S. Census boundary files (available from `http://www.census.gov/geo/www/cob/index.html`) are available in both proprietary and ASCII formats. An example ASCII file describing the counties of Massachusetts is available on the book Web site. The first few lines are reproduced here.

```
        1      -0.709816806854972E+02      0.427749187746914E+02
   -0.709148990000000E+02       0.428865890000000E+02
   -0.709148860000000E+02       0.428865640000000E+02
   -0.709148860000000E+02       0.428865640000000E+02
   -0.709027680000000E+02       0.428865300000000E+02
   -0.708861360000000E+02       0.428826100000000E+02
   -0.708837340828846E+02       0.428812223551543E+02
...
   -0.709148990000000E+02       0.428865890000000E+02
END
```

The first line contains an identifier for the county (linked with a county name in an additional file) and a latitude and longitude centroid within the polygon representing the county defined by the remaining points. The remaining points on the boundary do not contain the identifier. After the lines with the points, a line containing the word "END" is included. In addition, the county boundaries contain different numbers of points.

7.4.2 Read input files

Reading this kind of data requires some care in programming. We begin by reading in all of the input lines, keeping track of how many counties have been observed (based on how many lines include END). This information is needed for housekeeping purposes when collecting map points for each county.

```
# read in the data
openurl = url("http://www.math.smith.edu/r/data/co25_d00.dat")
input = readLines(openurl)
# figure out how many counties, and how many entries
num = length(grep("END", input))
allvals = length(input)
numentries = allvals-num
# create vectors to store data
county = numeric(numentries); lat = numeric(numentries)
long = numeric(numentries)
```

```
curval = 0   # number of counties seen so far
# loop through each line
for (i in 1:allvals) {
   if (input[i]=="END") {
      curval = curval + 1
   } else {
      # remove extraneous spaces
      nospace = gsub("[ ]+", " ", input[i])
      # remove space in first column
      nospace = gsub("^ ", "", nospace)
      splitstring = as.numeric(strsplit(nospace, " ")[[1]])
      len = length(splitstring)
      if (len==3) {  # new county
         curcounty = splitstring[1]
         county[i-curval] = curcounty
         lat[i-curval] = splitstring[2]
         long[i-curval] = splitstring[3]
      } else if (len==2) { # continue current county
         county[i-curval] = curcounty
         lat[i-curval] = splitstring[1]
         long[i-curval] = splitstring[2]
      }
   }
}
```

Each line of the input file is processed in turn. The `strsplit()` function is used to parse the input file. Lines containing END require incrementing the count of counties seen to date. If the line indicates the start of a new county, the new county number is saved. If the line contains 2 fields (another set of latitudes and longitudes), then this information is stored in the appropriate index (`i-curval`) of the output vectors.

Next we read in a dataset of county names. Later we will plot the Massachusetts counties, and annotate the plot with the names of the counties.

```
# read county names
countynames =
   read.table("http://www.math.smith.edu/r/data/co25_d00a.dat",
   header=FALSE)
names(countynames) = c("county", "countyname")
```

7.4.3   Plotting maps

To create the map, we begin by determining the plotting region, creating the plot of boundaries, then adding the county names at the internal point that was provided.

```
xvals = c(min(lat), max(lat))
yvals = c(range(long))
pdf("massachusettsmap.pdf")
plot(xvals, yvals, pch=" ", xlab="", ylab="", xaxt="n", yaxt="n")
counties = unique(county)
for (i in 1:length(counties)) {
   # first element is an internal point
   polygon(lat[county==counties[i]][-1],
      long[county==counties[i]][-1])
   # plot name of county using internal point
   text(lat[county==counties[i]][1],
      long[county==counties[i]][1], countynames$countyname[i])
}
dev.off()
```

Since the first set of points is in the interior of the county, these are not included in the values given to the polygon function (see indexing, Section 1.5.2).

The `pdf()` function is used to create an external graphics file (see 6.4.1, creating PDF files). When all plotting commands are complete, the `dev.off()` function (6.4.1) is used to close the graphics device. We display the results in Figure 7.3.

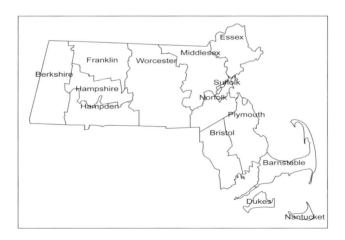

Figure 7.3: Massachusetts counties.

Many other mapping tools that support multiple projections are available in the `maps` package (see also the CRAN Spatial Statistics Task View).

7.5 Data scraping and visualization

In addition to the analytic capabilities available within R, the language has the capability of text processing. In the next sections, we automate data harvesting from the Web, by "scraping" a URL, then reading a datafile with two lines per observation, and plotting the results as time series data. The data being harvested and displayed are the sales ranks from Amazon for the *Manga Guide to Statistics* [68].

7.5.1 Scraping data from HTML files

We can find the Amazon Bestsellers Rank for a book by downloading the desired Web page and ferreting out the appropriate line. This code is highly sensitive to changes in Amazon's page format is changed (but it worked as of January, 2010). The code relies heavily on Section 2.1.4 (reading more complex data files) as well as Section 2.4.9 (replacing strings). To help in comprehending the code, readers are encouraged to run the commands on a line-by-line basis, then look at the resulting value.

```
# grab contents of web page
urlcontents = readLines("http://tinyurl.com/statsmanga")
# find line with bestsellers rank
linenum = suppressWarnings(grep("Amazon Bestsellers Rank:",
            urlcontents))
# split line into multiple elements
linevals = strsplit(urlcontents[linenum], ' ')[[1]]
# find element with bestsellers rank number
entry = grep("#", linevals)
# snag that entry
charrank = linevals[entry]
# kill '#' at start
charrank = substr(charrank, 2, nchar(charrank))
charrank = gsub(',','', charrank)   # remove commas
# turn it into a numeric object
bestsellersrank = as.numeric(charrank)
cat("bestsellersrank=",bestsellersrank,"\n")
```

7.5.2 Reading data with two lines per observation

The code from the previous entry was run regularly on a server by calling R in batch mode (see 1.2.2), with entries stored in a file. While a date-stamp was added when this was called, unfortunately it was included in the output on a different line. The file (accessible at http://www.math.smith.edu/r/data/manga.txt) has the following form.

```
Thu Dec 31 03:40:03 EST 2009
salesrank= 30531
Thu Dec 31 04:00:03 EST 2009
salesrank= 31181
```

We begin by reading the file, then calculate the number of entries by dividing the file's length by two. Next, two empty vectors of the correct length and type are created to store the data. Once this preparatory work is completed, we loop (2.11.1) through the file, reading in the odd-numbered lines as date/time values from the Eastern U.S. time zone, with daylight savings applied. The gsub() function (2.4.9) replaces matches determined by regular expression matching. In this situation, it is used to remove the time zone from the line before this processing. These date/time values are read into the timeval vector. Even-numbered lines are read into the rank vector, after removing the strings salesrank= and NA (again using two calls to gsub()). Finally, we make a data frame (1.5.6) from the two vectors and display the first few lines using the head() function (2.2.7).

```
file = readLines("manga.txt")
n = length(file)/2
rank = numeric(n)
timeval = as.POSIXlt(rank, origin="1960-01-01")
for (i in 1:n) {
   timeval[i] = as.POSIXlt(gsub('EST', '',
      gsub('EDT', '', file[(i-1)*2+1])),
      tz="EST5EDT", format="%a %b %d %H:%M:%S %Y")
   rank[i] = as.numeric(gsub('NA', '',
      gsub('salesrank= ','', file[i*2])))
}
timerank = data.frame(timeval, rank)
```

The first 7 entries of the file are given below.

```
> head(timerank, 7)
                timeval  rank
1  2009-12-30 07:14:27 18644
2  2009-12-30 07:20:03 18644
3  2009-12-30 07:40:03 18644
4  2009-12-30 08:00:03 18906
5  2009-12-30 08:20:02 18906
6  2009-12-30 08:40:03 18906
7  2009-12-30 09:00:04 13126
```

7.5.3   Plotting time series data

While it is straightforward to plot these data using the `plot()` command (6.1.1),
we can augment the display by indicating whether the rank was recorded in
nighttime (Eastern U.S. time) or not. Then we will color (6.3.11) the nighttime
ranks differently than the daytime ranks.

We begin by creating a new variable reflecting the date-time at the midnight
before we started collecting data. We then coerce the time values to numeric
values using the `as.numeric()` function (2.4.2) while subtracting that midnight
value. Next, we mod by 24 (using the `%%` operator, Section 1.5.3) and lastly
round to the integer value (2.8.4) to get the hour of measurement.

```
midnight = as.POSIX1t("2009-12-30 00:00:00 EST")
timeofday = round(as.numeric(timeval-midnight)%%24,0)
night = rep(0,length(timeofday))  # vector of zeroes
night[timeofday < 8 | timeofday > 18] = 1
```

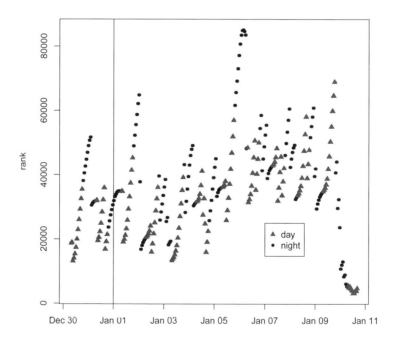

Figure 7.4: Plot of sales rank over time.

```
plot(timeval, rank, type="n")
points(timeval[night==1], rank[night==1], pch=20, col="black")
points(timeval[night==0], rank[night==0], pch=17, col="red")
legend(as.POSIXlt("2010-01-07 00:00:00 EST"), 25000,
    legend=c("day","night"), col=c("red","black"), pch=c(17,20))
abline(v=as.numeric(as.POSIXlt("2010-01-01 00:00:00 EST")))
```

The results are displayed in Figure 7.4, with a line denoting the start of the New Year. The sales rank gradually increases in a steady manner most of the time, then drops considerably when there is a sale.

7.6 Account for missing data using multiple imputation

Missing data is ubiquitous in most real-world investigations. Here we demonstrate some of the capabilities for fitting incomplete data regression models using multiple imputation [56, 63, 27] implemented with chained equation models [75, 50].

In this example we replicate an analysis from Section 5.7.1 in a version of the HELP dataset that includes missing values for several of the predictors. While not part of the regression model of interest, the mcs and pcs variables are included in the imputation models, which may make the missing at random assumption more plausible [6].

We begin by reading in the data then using the na.pattern() function from the Hmisc library to characterize the patterns of missing values within the dataframe.

```
> ds = read.csv("http://www.math.smith.edu/r/data/helpmiss.csv")
> smallds = with(ds, data.frame(homeless, female, i1, sexrisk,
    indtot, mcs, pcs))
```

```
> summary(smallds)
    homeless           female              i1              sexrisk
 Min.   :0.000    Min.   :0.000    Min.    :  0.0    Min.    : 0.00
 1st Qu.:0.000    1st Qu.:0.000    1st Qu.:  3.0    1st Qu.: 3.00
 Median :0.000    Median :0.000    Median : 13.0    Median : 4.00
 Mean   :0.466    Mean   :0.236    Mean    : 18.3    Mean    : 4.64
 3rd Qu.:1.000    3rd Qu.:0.000    3rd Qu.: 26.0    3rd Qu.: 6.00
 Max.   :1.000    Max.   :1.000    Max.    :142.0    Max.    :14.00
                                                      NA's    : 1.00

     indtot            mcs              pcs
 Min.   : 4.0    Min.    : 6.76    Min.   :14.1
 1st Qu.:32.0    1st Qu.:21.66    1st Qu.:40.3
 Median :37.5    Median :28.56    Median :48.9
 Mean   :35.7    Mean    :31.55    Mean   :48.1
 3rd Qu.:41.0    3rd Qu.:40.64    3rd Qu.:57.0
 Max.   :45.0    Max.    :62.18    Max.   :74.8
 NA's   :14.0    NA's    : 2.00    NA's   : 2.0
> library(Hmisc)
> na.pattern(smallds)
pattern
0000000 0000011 0000100 0001100
    454       2      13       1
```

There are 14 subjects missing `indtot`, 2 missing `mcs` and `pcs`, and 1 missing `sexrisk`. In terms of patterns of missingness, there are 454 observations with complete data, 2 missing both `mcs` and `pcs`, 13 missing `indtot` alone, and 1 missing `sexrisk` and `indtot`. Fitting a logistic regression model (5.1.1) using the available data (n=456) yields the following results.

```
> glm(homeless ~ female + i1 + sexrisk + indtot, binomial,
      data=smallds)
Call:  glm(formula = homeless ~ female + i1 + sexrisk + indtot,
    family = binomial, data = smallds)


Coefficients:
(Intercept)       female            i1         sexrisk        indtot
    -2.5278       -0.2401        0.0232          0.0562        0.0493

(14 observations deleted due to missingness)
```

Next, the `mice()` function within the `mice` library is used to impute missing values for `sexrisk`, `indtot`, `mcs`, and `pcs`. These results are combined using `glm.mids()`, and results are pooled and reported. Note that by default, all variables within the `smallds` data frame are included in each of the chained equations (e.g., `mcs` and `pcs` are used as predictors in each of the imputation models).

```
> library(mice)
> imp = mice(smallds, m=25, maxit=25, seed=42)

> summary(pool(glm.mids(homeless ~ female + i1 + sexrisk +
    indtot, family=binomial, data=imp)))
                 est      se      t  df Pr(>|t|)     lo 95    hi 95
(Intercept) -2.5366 0.59460 -4.266 456 2.42e-05  -3.7050  -1.3681
female      -0.2437 0.24393 -0.999 464 3.18e-01  -0.7230   0.2357
i1           0.0231 0.00561  4.114 464 4.61e-05   0.0121   0.0341
sexrisk      0.0590 0.03581  1.647 463 1.00e-01  -0.0114   0.1294
indtot       0.0491 0.01582  3.105 455 2.02e-03   0.0180   0.0802
            missing     fmi
(Intercept)      NA 0.01478
female            0 0.00182
i1                0 0.00143
sexrisk           1 0.00451
indtot           14 0.01728
```

The summary includes the number of missing observations as well as the fraction of missing information (fmi). While the results are qualitatively similar, they do differ, which is not surprising given the different imputation models used.

Support for other missing data models is available in the `mix` and `mitools` packages.

7.7 Propensity score modeling

Propensity scores can be used to attempt to approximate a randomized setting in an observational setting where there are potential confounding factors [54, 55]. Here we consider comparisons of the PCS scores for homeless vs. non-homeless subjects in the HELP study. Clearly, subjects were not randomized to homelessness, so if we want to make causal inference about the effects of homelessness, we need to adjust to selection bias with respect to homeless status (as the homeless subjects may systematically different from the non-homeless on other factors).

First, we examine the relationship between homelessness and PCS (physical component score) without adjustment.

```
> ds = read.csv("http://www.math.smith.edu/r/data/help.csv")
> attach(ds)
> lm1 = lm(pcs ~ homeless)
> summary(lm1)
Call:
lm(formula = pcs ~ homeless)

Residuals:
    Min       1Q   Median       3Q      Max
-34.9265  -7.9030   0.6438   8.3869  25.8055

Coefficients:
            Estimate Std. Error t value Pr(>|t|)
(Intercept)   49.001      0.688  71.220   <2e-16 ***
homeless      -2.064      1.013  -2.038   0.0422 *

Residual standard error: 10.75 on 451 degrees of freedom
Multiple R-squared: 0.009123,Adjusted R-squared: 0.006926
F-statistic: 4.152 on 1 and 451 DF,  p-value: 0.04216
```

We see a statistically significant difference in PCS scores ($p = 0.042$), but this unadjusted comparison does not account for other differences in homeless status, and thus may be confounded.

One approach to this problem involves controlling for possible confounders (in this case, age, gender, number of drinks, and MCS score) in a multiple regression model (4.1.1). This yields the following results.

```
> lm2 = lm(pcs ~ homeless + age + female + i1 + mcs)
> summary(lm2)
Call:
lm(formula = pcs ~ homeless + age + female + i1 + mcs)

Coefficients:
            Estimate Std. Error t value Pr(>|t|)
(Intercept) 58.21224    2.56675  22.679  < 2e-16 ***
homeless    -1.14707    0.99794  -1.149 0.250992
age         -0.26593    0.06410  -4.148 4.01e-05 ***
female      -3.95519    1.15142  -3.435 0.000648 ***
i1          -0.08079    0.02538  -3.184 0.001557 **
mcs          0.07032    0.03807   1.847 0.065396 .

Residual standard error: 10.22 on 447 degrees of freedom
Multiple R-squared: 0.1117,Adjusted R-squared: 0.1017
F-statistic: 11.24 on 5 and 447 DF,  p-value: 3.209e-10
```

Controlling for the other predictors has caused the parameter estimate to attenuate to the point that it is no longer statistically significant ($p = 0.25$). While controlling for other confounders appears to be effective in this problem, other situations may be more vexing, particularly if the dataset is small and the number of measured confounders is large. In such settings, estimation of the propensity score (the probability of being homeless, conditional on other factors), can be used to account for the set of unmeasured confounders, or can be used to match comparable subjects based on values of the propensity score.

Estimation of the propensity score is straightforward using a logistic regression model (5.1.1). A `formula` object (see 4.1.1) is used to specify the model.

```
> form = formula(homeless ~ age + female + i1 + mcs)
> glm1 = glm(form, family=binomial)
> X = glm1$fitted
> lm3 = lm(pcs ~ homeless + X)
```

```
> summary(lm3)

Call:
lm(formula = pcs ~ homeless + X)

Residuals:
     Min       1Q   Median       3Q      Max
-34.0278  -7.6229   0.9298   8.2433  25.6487

Coefficients:
             Estimate Std. Error t value Pr(>|t|)
(Intercept)    54.539      1.825  29.880  < 2e-16 ***
homeless       -1.178      1.038  -1.135  0.25690
X             -12.889      3.942  -3.270  0.00116 **

Residual standard error: 10.63 on 450 degrees of freedom
Multiple R-squared: 0.03212, Adjusted R-squared: 0.02782
F-statistic: 7.468 on 2 and 450 DF,  p-value: 0.000645
```

As with the multiple regression model, controlling for the propensity also leads to an attenuated estimate of the homeless coefficient, though this model only requires 1 degree of freedom in the regression model of interest.

Another approach uses the propensity score as a tool to create a matched sample that is relatively balanced on the terms included in the propensity model. This is straightforward to do using the `Matching` library.

```
> library(Matching)
> rr = Match(Y=pcs, Tr=homeless, X=X, M=1)
> summary(rr)

Estimate...   -0.80207
AI SE......    1.4448
T-stat.....   -0.55516
p.val......    0.57878


Original number of observations..............  453
Original number of treated obs...............  209
Matched number of observations...............  209
Matched number of observations (unweighted).  252
```

We see that the causal estimate of -0.80 in the matched comparison is not statistically significant ($p = 0.58$), which is similar to the other approaches that accounted for the confounders. The `MatchBalance()` function can be used to describe the distribution of the predictors (by homeless status) before and after matching (to save space, only the results for `age` and `i1` are displayed).

```
> MatchBalance(form, match.out=rr, nboots=10)
***** (V1) age *****   Before Matching     After Matching
mean treatment........     36.368            36.368
mean control..........     35.041            36.423
std mean diff.........     16.069           -0.65642

mean raw eQQ diff.....     1.5981            0.94841
med  raw eQQ diff.....          1                 1
max  raw eQQ diff.....          7                10

mean eCDF diff........   0.037112           0.022581
med  eCDF diff........   0.026365           0.019841
max  eCDF diff........    0.10477           0.083333

var ratio (Tr/Co).....     1.3290            1.2671
T-test p-value........   0.070785            0.93902
KS Bootstrap p-value.. < 2.22e-16               0.3
KS Naive p-value......    0.16881            0.34573
KS Statistic..........    0.10477           0.083333
```

```
***** (V3) i1 *****      Before Matching      After Matching
mean treatment........       23.038               23.038
mean control..........       13.512               20.939
std mean diff.........       40.582                8.945

mean raw eQQ diff.....        9.6316               2.1071
med  raw eQQ diff.....        8                    1
max  raw eQQ diff.....       73                   66

mean eCDF diff........        0.11853              0.018753
med  eCDF diff........        0.12377              0.011905
max  eCDF diff........        0.20662              0.087302

var ratio (Tr/Co).....        2.3763               1.3729
T-test p-value........  7.8894e-07                 0.011786
KS Bootstrap p-value..  < 2.22e-16                 0.3
KS Naive p-value......  0.00013379                 0.29213
KS Statistic..........        0.20662              0.087302
```

Both these variables had distributions that were considerably closer to each other in the matched sample than in the original dataset.

The Match() function can also be used to generate a dataset containing only the matched observations (see the index.treated and index.control components of the return value).

7.8 Empirical problem solving

7.8.1 Diploma (or hat-check) problem

Smith College is a residential women's liberal arts college in Northampton, MA that is steeped in tradition. One such tradition is to give every graduating student a diploma at random (or more accurately, in a haphazard fashion). At the end of the ceremony, a circle is formed, and students repeatedly pass the diplomas to the person next to them, stepping out once they have received their own diploma. This problem, also known as the *hat-check* problem, is featured in Mosteller [44]. Variants provide great fodder for probability courses.

The analytic (closed-form) solution for the expected number of students who receive their diplomas in the initial disbursement is very straightforward. Let X_i be the event that the ith student receives their diploma. $E[X_i] = 1/n$ for all i, since the diplomas are assumed uniformly distributed. If T is defined as the sum of all of the events X_1 through X_n, $E[T] = n*1/n = 1$ by the rules of expectations. It is sometimes surprising to students that this result does not depend on n. The variance is trickier, since the outcomes are not independent (if the first student receives their diploma, the probability that the others will increases ever so slightly).

For students, the use of empirical (simulation-based) problem solving is increasingly common as a means to complement and enhance analytic (closed-form) solutions. Here we illustrate how to simulate the expected number of students who receive their diploma as well as the standard deviation of that quantity. We assume that $n = 650$.

We begin by setting up some constants and a vector for results that we will use to store results. The students vector can be generated once, with the permuted vector of diplomas generated inside the loop generated using sample() (see 2.5.2). The == operator (1.5.2) is used to compare each of the elements of the vectors.

```
numsim = 100000
n = 650
res = numeric(numsim)
students = 1:n
for (i in 1:numsim) {
    diploma = sample(students, n)
    res[i] = sum(students==diploma)
}
```

This generates the following output.

```
> table(res)
    0       1       2       3       4       5       6       7       8
36568   36866   18545   6085    1590    295     40      9       2
> mean(res)
[1] 1.00365
> sd(res)
[1] 0.9995232
```

The expected value and standard deviation of the number of students who receive their diplomas on the first try are both 1.

7.8.2   Knapsack problem (constrained optimization)

The Web site http://rosettacode.org/wiki/Knapsack_Problem describes a fanciful trip by a traveler to Shangri-La. Upon leaving, they are allowed to take as much of three valuable items as they like, as long as they fit in a knapsack. A maximum of 25 weights can be taken, with a total volume of 25 cubic units. The weights, volumes, and values of the three items are given in Table 7.1.

It is straightforward to calculate the solutions using brute force, by iterating over all possible combinations and eliminating those that are over weight or too large to fit. A number of support functions are defined, then run over all possible values of the knapsack contents (after expand.grid() generates the list). The findvalue() function checks the constraints and sets the value to 0 if they are not satisfied, and otherwise calculates them for the set. The apply() function

Table 7.1: Weights, Volume, and Values for the Knapsack Problem

Item	Weight	Volume	Value
I	0.3	2.5	3000
II	0.2	1.5	1800
III	2.0	0.2	2500

(see 2.13.6) is used to run a function for each item of a vector.

```
# Define constants and useful functions
weight = c(0.3, 0.2, 2.0)
volume = c(2.5, 1.5, 0.2)
value = c(3000, 1800, 2500)
maxwt = 25
maxvol = 25


# minimize the grid points we need to calculate
max.items = floor(pmin(maxwt/weight, maxvol/volume))


# useful functions
getvalue = function(n) sum(n*value)
getweight = function(n) sum(n*weight)
getvolume = function(n) sum(n*volume)


# main function: return 0 if constraints not met,
# otherwise return the value of the contents, and their weight
findvalue = function(x) {
   thisweight = apply(x, 1, getweight)
   thisvolume = apply(x, 1, getvolume)
   fits = (thisweight <= maxwt) &
          (thisvolume <= maxvol)
   vals = apply(x, 1, getvalue)
   return(data.frame(I=x[,1], II=x[,2], III=x[,3],
       value=fits*vals, weight=thisweight,
       volume=thisvolume))
}


# Find and evaluate all possible combinations
combs = expand.grid(lapply(max.items, function(n) seq.int(0, n)))
values = findvalue(combs)
```

Now we can display the solutions.

```
> max(values$value)
[1] 54500
> values[values$value==max(values$value),]
     I II III value weight volume
2067 9  0  11 54500   24.7   24.7
2119 6  5  11 54500   24.8   24.7
2171 3 10  11 54500   24.9   24.7
2223 0 15  11 54500   25.0   24.7
```

The first solution (with 9 of item I), no item II, and 11 of item III satisfies the volume constraint, maximizes the value, and also minimizes the weight. More sophisticated approaches are available using the lpSolve package for linear/integer problems.

7.9   Further resources

Rubin's review [56] and Schafer's book [63] provide overviews of multiple imputation, while Van Buuren, Boshuizen, and Knook [75] and Raghunathan, Lepkowski, van Hoewyk, and Solenberge [50] describe chained equation models. Horton and Lipsitz [28] and Horton and Kleinman [27] provide a review of software implementations of missing data models.

Rizzo's text [52] provides a comprehensive review of statistical computing tasks implemented using R, while Horton, Brown, and Qian [25] describe the use of R as a toolbox for mathematical statistics exploration.

Appendix

The HELP study dataset

A.1 Background on the HELP study

Data from the HELP (Health Evaluation and Linkage to Primary Care) study are used to illustrate many of the entries. The HELP study was a clinical trial for adult inpatients recruited from a detoxification unit. Patients with no primary care physician were randomized to receive a multidisciplinary assessment and a brief motivational intervention or usual care, with the goal of linking them to primary medical care. Funding for the HELP study was provided by the National Institute on Alcohol Abuse and Alcoholism (R01-AA10870, Samet PI) and National Institute on Drug Abuse (R01-DA10019, Samet PI).

Eligible subjects were adults, who spoke Spanish or English, reported alcohol, heroin or cocaine as their first or second drug of choice, resided in proximity to the primary care clinic to which they would be referred or were homeless. Patients with established primary care relationships they planned to continue, significant dementia, specific plans to leave the Boston area that would prevent research participation, failure to provide contact information for tracking purposes, or pregnancy were excluded.

Subjects were interviewed at baseline during their detoxification stay and follow-up interviews were undertaken every 6 months for 2 years. A variety of continuous, count, discrete, and survival time predictors and outcomes were collected at each of these five occasions.

The details of the randomized controlled trial along with the results from a series of additional observational analyses have been published [59, 51, 29, 40, 34, 58, 57, 65, 35, 79].

A.2 Road map to analyses of the HELP dataset

Table A.1 summarizes the analyses illustrated using the HELP dataset. These analyses are intended to help illustrate the methods described in the book.

Interested readers are encouraged to review the published data from the HELP study for substantive analyses.

Table A.1: Analyses Illustrated Using the HELP Dataset

Description	Section
Data input and output	2.13.1
Summarize data contents	2.13.1
Data display	2.13.4
Derived variables and data manipulation	2.13.5
Sorting and subsetting	2.13.6
Summary statistics	3.6.1
Exploratory data analysis	3.6.1
Bivariate relationship	3.6.2
Contingency tables	3.6.3
Two-sample tests	3.6.4
Survival analysis (logrank test)	3.6.5
Scatterplot with smooth fit	4.7.1
Regression with prediction intervals	4.7.2
Linear regression with interaction	4.7.3
Regression diagnostics	4.7.4
Fitting stratified regression models	4.7.5
Two-way analysis of variance (ANOVA)	4.7.6
Multiple comparisons	4.7.7
Contrasts	4.7.8
Logistic regression	5.7.1
Poisson regression	5.7.2
Zero-inflated Poisson regression	5.7.3
Negative binomial regression	5.7.4
Lasso model selection	5.7.5
Quantile regression	5.7.6
Ordinal logit	5.7.7
Multinomial logit	5.7.8
Generalized additive model	5.7.9
Reshaping datasets	5.7.10
General linear model for correlated data	5.7.11
Random effects model	5.7.12
Generalized estimating equations model	5.7.13
Generalized linear mixed model	5.7.14
Proportional hazards regression model	5.7.15
Bayesian Poisson regression	5.7.16
Cronbach α	5.7.17
Factor analysis	5.7.18
Recursive partitioning	5.7.20

Linear discriminant analysis	5.7.21
Hierarchical clustering	5.7.22
Scatterplot with multiple y axes	6.6.1
Bubbleplot	6.6.2
Conditioning plot	6.6.3
Multiple plots	6.6.4
Dotplot	6.6.5
Kaplan–Meier plot	6.6.6
ROC curve	6.6.7
Pairs plot	6.6.8
Visualize correlation matrix	6.6.9
Multiple imputation	7.6
Propensity score modeling	7.7

A.3 Detailed description of the dataset

The Institutional Review Board of Boston University Medical Center approved all aspects of the study, including the creation of the de-identified dataset. Additional privacy protection was secured by the issuance of a Certificate of Confidentiality by the Department of Health and Human Services.

A de-identified dataset containing the variables utilized in the end of chapter examples is available for download at the book Web site:
`http://www.math.smith.edu/r/data/help.csv`
Variables included in the HELP dataset are described in Table A.2. A copy of the study instruments can be found at: `http://www.math.smith.edu/help`

Table A.2: Annotated Description of Variables in the HELP Dataset

VARIABLE	DESCRIPTION (VALUES)	NOTE
a15a	number of nights in overnight shelter in past 6 months (range 0–180)	see also `homeless`
a15b	number of nights on the street in past 6 months (range 0–180)	see also `homeless`
age	age at baseline (in years) (range 19–60)	
anysubstatus	use of any substance postdetox (0=no, 1=yes)	see also `daysanysub`
cesd*	Center for Epidemiologic Studies Depression scale (range 0–60)	see also `f1a–f1t`

d1	how many times hospitalized for medical problems (lifetime) (range 0–100)	
daysanysub	time (in days) to first use of any substance postdetox (range 0–268)	see also anysubstatus
daysdrink	time (in days) to first alcoholic drink post-detox (range 0–270)	see also drinkstatus
dayslink	time (in days) to linkage to primary care (range 0–456)	see also linkstatus
drinkstatus	use of alcohol postdetox (0=no, 1=yes)	see also daysdrink
drugrisk*	Risk-Assessment Battery (RAB) drug risk score (range 0–21)	see also sexrisk
e2b*	number of times in past 6 months entered a detox program (range 1–21)	
f1a	I was bothered by things that usually don't bother me (range 0–3#)	
f1b	I did not feel like eating; my appetite was poor (range 0–3#)	
f1c	I felt that I could not shake off the blues even with help from my family or friends (range 0–3#)	
f1d	I felt that I was just as good as other people (range 0–3#)	
f1e	I had trouble keeping my mind on what I was doing (range 0–3#)	
f1f	I felt depressed (range 0–3#)	
f1g	I felt that everything I did was an effort (range 0–3#)	
f1h	I felt hopeful about the future (range 0–3#)	
f1i	I thought my life had been a failure (range 0–3#)	
f1j	I felt fearful (range 0–3#)	
f1k	My sleep was restless (range 0–3#)	
f1l	I was happy (range 0–3#)	

f1m	I talked less than usual (range 0–3$^\#$)	
f1n	I felt lonely (range 0–3$^\#$)	
f1o	People were unfriendly (range 0–3$^\#$)	
f1p	I enjoyed life (range 0–3$^\#$)	
f1q	I had crying spells (range 0–3$^\#$)	
f1r	I felt sad (range 0–3$^\#$)	
f1s	I felt that people dislike me (range 0–3$^\#$)	
f1t	I could not get going (range 0–3$^\#$)	
female	gender of respondent (0=male, 1=female)	
g1b*	experienced serious thoughts of suicide (last 30 days, values 0=no, 1=yes)	
homeless*	1 or more nights on the street or shelter in past 6 months (0=no, 1=yes)	see also a15a and a15b
i1*	average number of drinks (standard units) consumed per day (in the past 30 days, range 0–142)	see also i2
i2	maximum number of drinks (standard units) consumed per day (in the past 30 days range 0–184)	see also i1
id	random subject identifier (range 1–470)	
indtot*	Inventory of Drug Use Consequences (InDUC) total score (range 4–45)	
linkstatus	post-detox linkage to primary care (0=no, 1=yes)	see also dayslink
mcs*	SF-36 Mental Component Score (range 7–62)	see also pcs
pcrec*	number of primary care visits in past 6 months (range 0–2)	see also linkstatus, not observed at baseline
pcs*	SF-36 Physical Component Score (range 14-75)	see also mcs

pss_fr	perceived social supports (friends, range 0–14)	see also `dayslink`
satreat	any BSAS substance abuse treatment at baseline (0=no, 1=yes)	
sexrisk*	Risk-Assessment Battery (RAB) drug risk score (range 0–21)	see also `drugrisk`
substance	primary substance of abuse (alcohol, cocaine or heroin)	
treat	randomization group (0=usual care, 1=HELP clinic)	

Notes: Observed range is provided (at baseline) for continuous variables.

* denotes variables measured at baseline and followup (e.g., cesd is baseline measure, cesd1 is measure at 6 months, and cesd4 is measure at 24 months).
#: For each of the 20 items in HELP section F1 (CESD), respondents were asked to indicate how often they behaved this way during the past week (0 = rarely or none of the time, less than 1 day; 1 = some or a little of the time, 1 to 2 days; 2 = occasionally or a moderate amount of time, 3 to 4 days; or 3 = most or all of the time, 5 to 7 days); items f1d, f1h, f1l, and f1p were reverse coded.

Bibliography

[1] A. Agresti. *Categorical Data Analysis*. New York: John Wiley & Sons, 2002.

[2] J. Albert. *Bayesian Computation with R*. New York: Springer, 2008.

[3] T. S. Breusch and A. R. Pagan. A simple test for heteroscedasticity and random coefficient variation. *Econometrica*, 47, 1979.

[4] D. Collett. *Modelling Binary Data*. New York: Chapman & Hall, 1991.

[5] D. Collett. *Modeling Survival Data in Medical Research*. (2nd ed.). Boca Raton, FL: Chapman & Hall/CRC Press, 2003.

[6] L. M. Collins, J. L. Schafer, and C. M. Kam. A comparison of inclusive and restrictive strategies in modern missing data procedures. *Psychological Methods*, 6(4):330–351, 2001.

[7] R. D. Cook. *Residuals and Influence in Regression*. New York: Chapman & Hall, 1982.

[8] A. Damico. Transitioning to R: Replicating SAS, Stata, and SUDAAN Analysis Techniques in Health Policy Data. *The R Journal*, 1(2):37–44, December 2009.

[9] A. J. Dobson and A. Barnett. *An Introduction to Generalized Linear Models* (3rd ed.). Boca Raton, FL: Chapman & Hall/CRC Press, 2008.

[10] W. D. Dupont and W. D. Plummer. Density distribution sunflower plots. *Journal of Statistical Software*, 8:1–11, 2003.

[11] B. Efron and R. J. Tibshirani. *An Introduction to the Bootstrap*. New York: Chapman & Hall, 1993.

[12] M. J. Evans and J. S. Rosenthal. *Probability and Statistics: The Science of Uncertainty*. New York: W. H. Freeman, 2004.

[13] B. S. Everitt and T. Hothorn. *A Handbook of Statistical Analyses Using R*. (2nd ed.). Boca Raton, FL: Chapman & Hall/CRC Press, 2009.

[14] J. J. Faraway. *Linear Models with R.* Boca Raton, FL: Chapman & Hall/CRC Press, 2004.

[15] J. J. Faraway. *Extending the Linear Model with R: Generalized Linear, Mixed Effects and Nonparametric Regression Models.* Boca Raton, FL: Chapman & Hall/CRC Press, 2005.

[16] N. I. Fisher. *Statistical Analysis of Circular Data.* Cambridge, UK: Cambridge University Press, 1996.

[17] G. M. Fitzmaurice, N. M. Laird, and J. H. Ware. *Applied Longitudinal Analysis.* New York: Wiley, 2004.

[18] T. R. Fleming and D. P. Harrington. *Counting Processes and Survival Analysis.* New York: John Wiley & Sons, 1991.

[19] J. Fox. The R Commander: A basic graphical user interface to R. *Journal of Statistical Software*, 14(9), 2005.

[20] J. Fox. Aspects of the social organization and trajectory of the R project. *The R Journal*, 1(2):5–13, December 2009.

[21] A. Gelman, J. B. Carlin, H. S. Stern, and D. B. Rubin. *Bayesian Data Analysis.* (2nd ed.). Boca Raton, FL: Chapman & Hall Press, 2004.

[22] P. I. Good. *Permutation Tests: A Practical Guide to Resampling Methods for Testing Hypotheses.* New York: Springer-Verlag, 1994.

[23] J. W. Hardin and J. M. Hilbe. *Generalized Estimating Equations.* Boca Raton, FL: Chapman & Hall/CRC Press, 2002.

[24] T.C. Hesterberg, D. S. Moore, S. Monaghan, A. Clipson, and R. Epstein. *Bootstrap Methods and Permutation Tests.* New York: W. H. Freeman, 2005.

[25] N. J. Horton, E. R. Brown, and L. Qian. Use of R as a toolbox for mathematical statistics exploration. *The American Statistician*, 58(4):343–357, 2004.

[26] N. J. Horton, E. Kim, and R. Saitz. A cautionary note regarding count models of alcohol consumption in randomized controlled trials. *BMC Medical Research Methodology*, 7(9), 2007.

[27] N. J. Horton and K. P. Kleinman. Much ado about nothing: A comparison of missing data methods and software to fit incomplete data regression models. *The American Statistician*, 61:79–90, 2007.

[28] N. J. Horton and S. R. Lipsitz. Multiple imputation in practice: Comparison of software packages for regression models with missing variables. *The American Statistician*, 55(3):244–254, 2001.

[29] N. J. Horton, R. Saitz, N. M. Laird, and J. H. Samet. A method for modeling utilization data from multiple sources: Application in a study of linkage to primary care. *Health Services and Outcomes Research Methodology*, 3:211–223, 2002.

[30] R. Ihaka and R. Gentleman. R: A language for data analysis and graphics. *Journal of Computational and Graphical Statistics*, 5(3):299–314, 1996.

[31] K. Imai, G. King, and O. Lau. *Zelig: Everyone's Statistical Software*, 2008. R package version 3.3-1.

[32] W. G. Jacoby. The dot plot: A graphical display for labeled quantitative values. *The Political Methodologist*, 14(1):6–14, 2006.

[33] S. R. Jammalamadaka and A. Sengupta. *Topics in Circular Statistics*. World Scientific, 2001.

[34] S. G. Kertesz, N. J. Horton, P. D. Friedmann, R. Saitz, and J. H. Samet. Slowing the revolving door: stabilization programs reduce homeless persons' substance use after detoxification. *Journal of Substance Abuse Treatment*, 24:197–207, 2003.

[35] M. J. Larson, R. Saitz, N. J. Horton, C. Lloyd-Travaglini, and J. H. Samet. Emergency department and hospital utilization among alcohol and drug-dependent detoxification patients without primary medical care. *American Journal of Drug and Alcohol Abuse*, 32:435–452, 2006.

[36] M. Lavine. *Introduction to Statistical Thought*. M. Lavine, 2005. http://www.math.umass.edu/∼lavine/Book/book.html.

[37] F. Leisch. Sweave: Dynamic generation of statistical reports using literate data analysis. In Wolfgang Härdle and Bernd Rönz, eds., *Compstat 2002—Proceedings in Computational Statistics*, pages 575–580. Heidelberg: Physica-Verlag, 2002. ISBN 3-7908-1517-9.

[38] R. Lenth and S. Højsgaard. Literate programming in statistics with multiple languages. In revision, 2010.

[39] K. -Y. Liang and S. L. Zeger. Longitudinal data analysis using generalized linear models. *Biometrika*, 73:13–22, 1986.

[40] J. Liebschutz, J. B. Savetsky, R. Saitz, N. J. Horton, C. Lloyd-Travaglini, and J. H. Samet. The relationship between sexual and physical abuse and substance abuse consequences. *Journal of Substance Abuse Treatment*, 22(3):121–128, 2002.

[41] S. R. Lipsitz, N. M. Laird, and D. P. Harrington. Maximum likelihood regression methods for paired binary data. *Statistics in Medicine*, 9:1517–1525, 1990.

[42] B. F. J. Manly, *Multivariate Statistical Methods: A Primer* (3rd ed.). Boca Raton, FL: Chapman & Hall/CRC Press, 2004

[43] P. McCullagh and J. A. Nelder. *Generalized Linear Models*. New York: Chapman & Hall, 1989.

[44] F. Mosteller. *Fifty Challenging Problems in Probability with Solutions*. Mineola, NY: Dover Publications, 1987.

[45] P. Murrell. *R Graphics*. Boca Raton, FL: Chapman & Hall/CRC Press, 2005.

[46] P. Murrell. *Introduction to Data Technologies*. Boca Raton, FL: Chapman & Hall/CRC Press, 2009.

[47] N. J. D. Nagelkerke. A note on a general definition of the coefficient of determination. *Biometrika*, 78(3):691–692, 1991.

[48] National Institute of Alcohol Abuse and Alcoholism, Bethesda, MD. *Helping Patients Who Drink Too Much*, 2005.

[49] R Development Core Team. *R: A Language and Environment for Statistical Computing*. R Foundation for Statistical Computing, Vienna, Austria, 2008. ISBN 3-900051-07-0.

[50] T. E. Raghunathan, J. M. Lepkowski, J. van Hoewyk, and P. Solenberger. A multivariate technique for multiply imputing missing values using a sequence of regression models. *Survey Methodology*, 27(1):85–95, 2001.

[51] V. W. Rees, R. Saitz, N. J. Horton, and J. H. Samet. Association of alcohol consumption with HIV sex- and drug-risk behaviors among drug users. *Journal of Substance Abuse Treatment*, 21(3):129–134, 2001.

[52] M. L. Rizzo. *Statistical Computing with R*. Boca Raton, FL: Chapman & Hall/CRC Press, 2007.

[53] Joseph P. Romano and Andrew F. Siegel. *Counterexamples in probability and statistics*. Duxbury Press, 1986.

[54] P. R. Rosenbaum and D. B. Rubin. Reducing bias in observational studies using subclassification on the propensity score. *Journal of the American Statistical Association*, 79:516–524, 1984.

[55] P. R. Rosenbaum and D. B. Rubin. Constructing a control group using multivariate matched sampling methods that incorporate the propensity score. *The American Statistician*, 39:33–38, 1985.

[56] D. B. Rubin. Multiple imputation after 18+ years. *Journal of the American Statistical Association*, 91:473–489, 1996.

[57] R. Saitz, N. J. Horton, M. J. Larson, M. Winter, and J. H. Samet. Primary medical care and reductions in addiction severity: A prospective cohort study. *Addiction*, 100(1):70–78, 2005.

[58] R. Saitz, M. J. Larson, N. J. Horton, M. Winter, and J. H. Samet. Linkage with primary medical care in a prospective cohort of adults with addictions in inpatient detoxification: Room for improvement. *Health Services Research*, 39(3):587–606, 2004.

[59] J. H. Samet, M. J. Larson, N. J. Horton, K. Doyle, M. Winter, and R. Saitz. Linking alcohol- and drug-dependent adults to primary medical care: A randomized controlled trial of a multidisciplinary health intervention in a detoxification unit. *Addiction*, 98(4):509–516, 2003.

[60] J. M. Sarabia, E. Castillo, and D. J. Slottje. An ordered family of Lorenz curves. *Journal of Econometrics*, 91:43–60, 1999.

[61] D. Sarkar. *Lattice: Multivariate Data Visualization With R.* New York: Springer, 2008.

[62] C. -E. Särndal, B. Swensson, and J. Wretman. *Model Assisted Survey Sampling.* New York: Springer-Verlag, 1992.

[63] J. L. Schafer. *Analysis of Incomplete Multivariate Data.* New York: Chapman & Hall, 1997.

[64] G. A. F. Seber and C. J. Wild. *Nonlinear Regression.* New York: Wiley, 1989.

[65] C. W. Shanahan, A. Lincoln, N. J. Horton, R. Saitz, M. J. Larson, and J. H. Samet. Relationship of depressive symptoms and mental health functioning to repeat detoxification. *Journal of Substance Abuse Treatment*, 29:117–123, 2005.

[66] T. Sing, O. Sander, N. Beerenwinkel, and T. Lengauer. ROCR: Visualizing classifier performance in R. *Bioinformatics*, 21(20):3940–3941, 2005.

[67] B. G. Tabachnick and L. S. Fidell. *Using Multivariate Statistics.* (5th ed.). Boston: Allyn & Bacon, 2007.

[68] S. Takahashi. *The Manga Guide to Statistics.* San Francisco, CA: No Starch Press, 2008.

[69] R. Tibshirani. Regression shrinkage and selection via the LASSO. *Journal of the Royal Statistical Society B*, 58(1), 1996.

[70] E. R. Tufte. *Envisioning Information.* Cheshire, CT: Graphics Press, 1990.

[71] E. R. Tufte. *Visual Explanations: Images and Quantities, Evidence and Narrative.* Cheshire, CT: Graphics Press, 1997.

[72] E. R. Tufte. *Visual Display of Quantitative Information.* (2nd ed.). Cheshire, CT: Graphics Press, 2001.

[73] E. R. Tufte. *Beautiful Evidence.* Cheshire, CT: Graphics Press, 2006.

[74] J. W. Tukey. *Exploratory Data Analysis.* Reading, MA: Addison Wesley, 1977.

[75] S. van Buuren, H. C. Boshuizen, and D. L. Knook. Multiple imputation of missing blood pressure covariates in survival analysis. *Statistics in Medicine*, 18:681–694, 1999.

[76] W. N. Venables, D. M. Smith, and the R Development Core Team. An introduction to R: Notes on R: A programming environment for data analysis and graphics, version 2.8.0. http://cran.r-project.org/doc/manuals/R-intro.pdf, accessed January 1, 2009.

[77] J. Verzani. *Using R For Introductory Statistics.* Chapman & Hall/CRC, 2005.

[78] B. West, K. B. Welch, and A. T. Galecki. *Linear Mixed Models: A Practical Guide Using Statistical Software.* Boca Raton, FL: Chapman & Hall/CRC Press, 2006.

[79] J. D. Wines Jr., R. Saitz, N. J. Horton, C. Lloyd-Travaglini, and J. H. Samet. Overdose after detoxification: A prospective study. *Drug and Alcohol Dependence*, 89:161–169, 2007.

[80] D. Zamar, B. McNeney, and J. Graham. elrm: Software implementing exact-like inference for logistic regression models. *Journal of Statistical Software*, 21(3):1–18, 9 2007.

Indices

Separate indices are provided for subject (concept or task) and for commands. The primary entry for a command is given in **bold**. References to the examples and case studies are denoted in *italics*.

Subject index

References to the HELP examples are denoted in *italics*. The primary entry for a command is given in **bold**.

R index

References to the HELP examples are denoted in *italics*. The primary entry for a command is given in **bold**.